PROJECT AIR FORCE

T0108975

Improving the Air Force Small-Business Performance Expectations Methodology

Nancy Y. Moore, Amy G. Cox, Clifford A. Grammich, Judith D. Mele

Prepared for the United States Air Force

For more information on this publication, visit www.rand.org/t/RR1545

Library of Congress Cataloging-in-Publication Data is available for this publication.
ISBN: 978-0-8330-9512-1

Published by the RAND Corporation, Santa Monica, Calif.
© Copyright 2017 RAND Corporation
RAND® is a registered trademark.

Support RAND
Make a tax-deductible charitable contribution at
www.rand.org/giving/contribute

www.rand.org

Preface

The federal government has long sought to increase use of small businesses through procurement preferences. Recent statutes and regulations have specified that at least 23 percent of all eligible dollars obligated by the federal government should be awarded to small businesses.

In recent years, the value of Air Force small-business use has been fairly stable, but at a level significantly below the 23-percent federal goal as a fraction of total Air Force obligations. To increase small-business use rates, the Air Force Office of Small Business Programs recently implemented a Performance Expectations process to set goals for increasing small-business use by its major commands (MAJCOMs) and Program Executive Offices (PEOs). By using this new method for setting Performance Expectations, the Air Force Office of Small Business Programs seeks to focus small-business utilization goals for specific MAJCOMs and PEOs. This method uses U.S. Department of Defense (DoD)/Air Force Maximum Practicable (MaxPrac) Opportunity tools and past purchasing trends to identify contracts with other-than-small businesses that might be shifted to small businesses.[1] Fiscal year 2013 was the first time the Air Force used such a focused, detailed approach to achieve its goals for small businesses.

The Air Force Office of Small Business Programs asked RAND Project AIR FORCE (PAF) to review its method for setting Performance Expectations and recommend possible improvements, giving special attention to small-business purchases by industry, Product or Service Code, budget category, and industry trends. This report details our assessment and recommendations. The analysis was conducted within PAF's Resource Management Program as part of a project entitled "Analytic Support of Air Force Small-Business Policy" to help the Air Force improve its strategy for achieving small-business objectives. The draft report, issued on January 8, 2016, was reviewed by formal peer reviewers and U.S. Air Force subject-matter experts.

This report should interest acquisition, contracting, and small-business personnel within the Air Force.

RAND Project AIR FORCE

RAND Project AIR FORCE (PAF), a division of the RAND Corporation, is the U.S. Air Force's federally funded research and development center for studies and analyses. PAF provides the Air Force with independent analyses of policy alternatives affecting the development, employment, combat readiness, and support of current and future air, space, and

---

[1] MaxPrac "is an analytic methodology which uses Federal Procurement Data System–Next Generation (FPDS-NG) acquisition data to identify potential opportunities for increased small-business participation in *unclassified* contract awards" (U.S. Department of Defense Office of Small Business Programs, undated b).

cyber forces. Research is conducted in four programs: Force Modernization and Employment; Manpower, Personnel, and Training; Resource Management; and Strategy and Doctrine. The research reported here was prepared under contract FA7014-06-C-0001.

Additional information is available on the PAF website: http://www.rand.org/paf

Contents

Preface... iii

Figures.. vii

Tables.. ix

Summary xi

1. Introduction.. 1

    Background and Purpose 1

    Organization of This Report 2

2. Current Performance Expectations Methodology.. 3

3. Air Force Small-Business Spending by MAJCOM and PEO...................................... 11

4. Air Force Small-Business Spending by Budget Category and Industry...................... 19

5. Refining the Performance Expectations Methodology 31

6. Conclusions... 41

Appendix 43

Acknowledgments... 45

Abbreviations.. 47

References 49

Figures

S.1. Current Performance Expectations Methodology xii

S.2. Proposed Performance Expectations Methodology xv

2.1. Current Performance Expectations Methodology 6

3.1. Performance Expectations Methodology Needs to Account for Wide Variation Across
MAJCOMs 12

3.2. Nearly All PEOs Have Low Small-Business Use Rates 16

5.1. Proposed Performance Expectations Methodology 33

Tables

2.1. Steps in Current Performance Expectations Methodology 4

2.2. Example of Performance Expectations Methodology Categorization and Calculations
for a Notional MAJCOM/PEO 7

3.1. Total Air Force Goaling Dollars Spent by MAJCOM, Overall and with Small
Businesses, FY 2013 11

3.2. Little Change from FY 2012 to FY 2013 in Small-Business Performance Across
MAJCOMs 13

3.3. Total Air Force Goaling Dollars Spent by PEO, Overall and with Small Businesses,
FY 2013 15

3.4. PEOs Have Relatively Few Small-Business Dollars 17

4.1. Small-Business Use by Budget Category Varies Across DoD 20

4.2. Air Force Small-Business Use Varies by Budget Category Within a Single Industry 22

4.3. Many NAICS/PSC Groups Are in One Budget Category, But 14 Percent of Groups
Have High Spend and Multiple Categories 24

4.4. Small-Business Use Is at or Near Its Maximum in Many Industries 26

4.5. OTSBs Dominate Many Industries in Which the Air Force Makes Most of Its
Purchases 27

4.6. SAM Registrants, Including Small Firms, Outnumber Total Firms Enumerated in
Economic Census for Leading Air Force Industries 29

5.1. Performance Expectations Methodology and Proposed Revisions 33

5.2. Existing Data Allow Consideration of Many Elements in Performance
Expectations Methodology 35

5.3. Eliminating Saturated Markets Would Have the Greatest Effect on Performance
Expectations Calculations for MAJCOMs 36

5.4. Eliminating Saturated Markets and Using Army-Plus-Navy Comparison Would
Slightly Affect Performance Expectations Calculations for PEOs 40

Summary

The U.S. government is interested in helping small businesses, and it does so by funneling some of its purchasing to them by using procurement preferences. Its goal is to award small businesses 23 percent of all eligible federal obligations (Pub. L. 85-536, 2013). Federal agencies, including the U.S. Department of Defense (DoD), contribute toward this goal with agency-specific small-business goals. Within DoD, the military services have individual small-business goals that partly depend on the goods and services they purchase.[2] The U.S. Air Force has fallen short in recent years of the small-business improvement goals it set for itself. Its 2013 goal was 15.0 percent, but actual use was 14.5 percent; its 2012 goal was 16.99 percent, but actual use was 14.8 percent. In 2013, the Air Force initiated a new Performance Expectations initiative to increase small-business use by its major commands (MAJCOMs) and Program Executive Offices (PEOs), setting objectives for using small businesses tailored to spending patterns in each organization. The Air Force is trying to identify industries that have the most potential for increasing small-business expenditures, focusing especially on other-than-small business (OTSB) contracts that could be shifted to small businesses. The Air Force Office of Small Business Programs asked RAND Project AIR FORCE to review the new method for setting Performance Expectations objectives and to recommend possible improvements. This report responds to that request. The recommendations herein are meant to increase the methodology's accuracy and not to increase or decrease the Performance Expectations themselves.

How the Air Force's New Performance Expectations Methodology Works

The Air Force's current Performance Expectations methodology breaks out spending by "markets," which are defined as combinations of industries designated by six-digit North American Industry Classification System (NAICS) codes and goods designated by Product or Service Codes (PSCs). Both NAICS codes and PSCs are given on individual contract-action reports, and each NAICS/PSC combination specifies an individual market.[3] The combinations

---

[2] In 2013, the small-business goal for DoD was 22.5 percent. That goal was divided between the various services and defense agencies based on how much and what they buy (e.g., large weapons such as fighter airplanes, aircraft carriers and submarines and tanks versus food, clothing, small industrial items, and facilities support services) and how amenable their purchases may be to small business use. The Army goal was 26.5 percent; the Navy goal was 16.5 percent, and the Air Force goal was 15.0 percent (U.S. Department of Defense Office of Small Business Programs, undated c; U.S. Department of the Army, 2013; Stackley, 2012; McDade, 2013).

[3] NAICS codes indicate the industry within which the business is providing goods and services. PSCs indicate the type of good or service that is being purchased. PSCs are more specific indicators of goods and services than NAICS codes.

enable greater precision in the definition of buying markets than either metric would yield by itself.[4] The Air Force's new methodology is depicted in Figure S.1.

Figure S.1. Current Performance Expectations Methodology

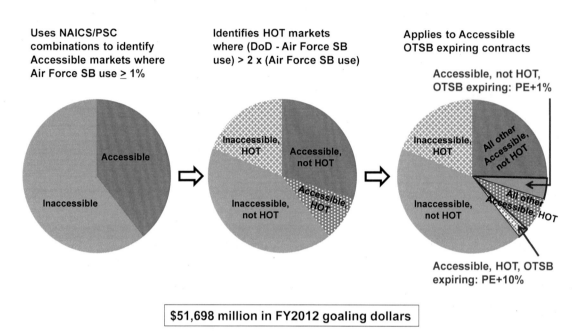

$51,698 million in FY2012 goaling dollars

NOTES: FY = fiscal year, HOT = High-Opportunity Target, PE = Performance Expectation, SB = small business.

Moving from left to right, the first step is to distinguish between Accessible and Inaccessible markets—i.e., those in which small-business contracts equal or exceed 1 percent and those in which they do not. The next step is to identify HOT markets, in which DoD small-business use (excluding the Air Force) is more than double that of the Air Force. HOT markets emphasize areas where the Air Force might increase its small-business contracts. The third step is to identify expiring OTSB contracts. The Air Force targets expiring contracts because it is ill advised to change contract awards midterm, and small businesses might be awarded expiring OTSB contracts. The fourth step is to identify which expiring OTSB contracts are in markets deemed Accessible HOT Spots, as well as all other Accessible expiring contracts (including small-business and OTSB contracts). The Air Force then identifies what each MAJCOM and PEO spent on small businesses in the past year, 10 percent of the dollars that are in all of its expiring OTSB contracts for Accessible HOT Spot markets, and 1 percent of the dollars that are in all of its expiring OTSB contracts for Accessible non–HOT Spot markets. Finally, the Air Force sums these three categories of dollars and divides them by what are known as "goaling dollars" spent

---

[4] Both NAICS codes and PSCs are fairly broad categories for the industries or goods/services that they define. Industries defined by NAICS codes can vary widely in size, manufacturing processes, customer type, and other characteristics.

in the past year to determine the Performance Expectation objective for the MAJCOM or PEO in the subsequent year.[5]

What the Air Force Methodology Shows About Small-Business Expenditures

Air Force MAJCOMs and PEOs vary in many ways (e.g., substantive area, size, location, supply chain), and some of their purchases may relate to small-business suppliers more directly than others. Applying the Air Force methodology to expenditures by MAJCOMs and PEOs reveals a number of trends. First, MAJCOMs and PEOs with the greatest spending have the smallest percentage of small business use (they primarily purchase large weapons that small businesses do not produce). For example, Air Force Materiel Command accounted for 71 and 73 percent of all Air Force goaling dollars in FYs 2012 and 2013, respectively, but only 11 percent of those dollars went to small businesses. PEOs spend more than 65 percent of Air Force goaling dollars, and most PEOs have relatively low small-business use rates, below 1 percent for four PEOs and between 1 and 5 percent for five others (out of 14 PEOs). While PEOs accounted for nearly two-thirds of all Air Force goaling dollars, they accounted for only about one-sixth of dollars that the Air Force spent with small businesses. Commands with smaller budgets typically exceeded both the overall Air Force goals (15 percent in 2013) and the federal goal of 23 percent for small-business use. Second, the level of small-business spending has not changed much over the past two FYs.

These analyses of spending demonstrate the need for separate small-business goals for these different components of the Air Force. The current Performance Expectations methodology deals with this variation effectively, and such accounting needs to be maintained through any refinements to the methodology.

An examination of expenditures by budget category shows that small-business use differs across the Air Force's budget categories, that the Air Force has a different distribution of spending by budget category than other parts of DoD, and that Air Force small-business use differs from that elsewhere in DoD. This finding raises questions about using DoD small-business spending as a whole as a standard for identifying HOT Spots for the Performance Expectations methodology. This analysis also underscores the need to account for budget categories when setting small-business performance expectations because not all goods and services are equally compatible with small-business procurement. Accounting for budget category offers a way to refine industry groupings and better identify HOT targets for increased small-business use.

---

[5] Goaling dollars exclude certain categories of expenditure, generally those that are not likely candidates for U.S. small-business contracts (e.g., contracts performed outside of the United States). For more on goaling exclusions, please see U.S. Department of Defense Office of Small Business Programs, undated a.

Analysis of expenditures by budget category by industry (aircraft manufacturing, engineering support services, and facilities support services) shows that considerable spending occurs outside the leading budget category for an industry and that small-business spending can differ between the leading and other categories in an industry. This finding suggests that some industries may be too broadly defined to easily identify areas ripe for small-business use improvement.

Findings and Recommendations

Our findings and recommendations include the following:

- Air Force spending by MAJCOM and PEO has been relatively consistent across time. Small-business use has often been greater for MAJCOMs and PEOs that spend fewer dollars and presumably have requirements more amenable to small-business purchases. The current Performance Expectations methodology recognizes this pattern, and any effort to refine it should do the same.
- Any efforts to increase small-business spending by MAJCOMs and PEOs should take into account differences among them, as well as differences with other services.
- Small-business spending differs by industry, PSC, and budget category. The current Performance Expectations methodology partially recognizes this by considering industry and PSC combinations, but it also needs to consider budget categories.
- Extant data can offer many insights but can be improved in some cases by narrowing industries that have multiple size standards or span multiple budget categories. Performance Expectations methodology addresses only one aspect of small-business use, determining and setting accurate goals for that use. Continuing to monitor and refine this methodology will help this tool remain useful as the Air Force seeks to increase its small-business use.

Figure S.2 illustrates the recommended changes to the Performance Expectations methodology to increase its accuracy. Key changes in the diagram include the following and are explained in more detail following the figure.

- Markets are defined as More Accessible or Less Accessible, rather than Accessible or Inaccessible.
- "DoD budget categories" is added to NAICS and PSC.
- "Saturated Markets" is added to the categories.
- Air Force small-business use is compared with Army and Navy combined use rather than DoD as a whole.

Figure S.2. Proposed Performance Expectations Methodology

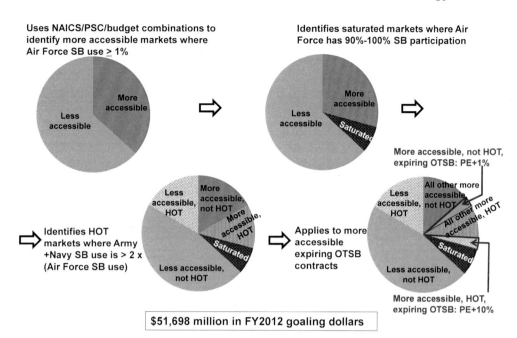

We recommend that Air Force use the terms More Accessible and Less Accessible to describe the markets, instead of Accessible and Inaccessible, to better define the opportunities in each category. We also recommend that the methodology use industry/PSC/budget-category groupings, rather than industry/PSC groupings, to identify More Accessible Markets. This addition will identify opportunities more precisely by better accounting for variation in small-business use by budget category. Within More Accessible Markets, we recommend that the Air Force use the category Saturated Markets, where the Air Force has at least 90 percent small-business use, making it difficult to expand small-business use.

We recommend that the Air Force identify certain expiring OTSB contracts as part of Consolidated Markets, in which small-business growth is less likely because market forces are causing firms either to increase in size or exit an industry. In Consolidated Markets, economies of scale or scope are increasing the size that a firm needs to compete, perhaps faster than Small Business Administration size standards recognize.

We also recommend that the Air Force identify expiring OTSB HOT Spot contracts by comparing its buying with that of the Army and Navy combined, or perhaps that of the Navy alone, but not with all of DoD. The services have more in common with each other in purchasing than they have with other elements of DoD. Recognizing this distinction will help the Air Force discern areas where it is most likely to be able to improve, as well as the most relevant lessons that others in DoD may offer.

Finally, we recommend that the Air Force subtract Saturated and OTSB expiring highly consolidated market contracts from OTSB expiring More Accessible not–HOT Spot contracts when calculating Performance Expectations. Eliminating these will help the Air Force set a more realistic and more targeted goal for the MAJCOMs and PEOs to achieve.

The Air Force has developed a useful methodology. The changes suggested here should improve an already useful tool by increasing its accuracy further and should help this tool remain useful as the Air Force seeks to increase its small-business use. The Air Force will also need to continue its ongoing efforts to examine other aspects of its purchasing that affect small-business use, including breaking out components from large weapon systems.

1. Introduction

Background and Purpose

For over 70 years, the federal government has sought to assist small businesses through procurement preferences. Recent statutes and regulations have specified that at least 23 percent of all eligible contract dollars obligated by the federal government as a whole should be awarded to small businesses (Pub. L. 85-536, 2013).[6] Federal agencies, including the U.S. Department of Defense (DoD), contribute toward this goal with agency-specific small-business goals. Within DoD, the military services have individual small-business goals that partly depend on the goods and services they purchase.[7]

In recent years, the value of Air Force small-business eligible contract dollars has remained fairly steady, ranging between 14 and 15 percent. To increase small-business use by its major commands (MAJCOMs) and Program Executive Offices (PEOs), the Air Force implemented a new Performance Expectations initiative in 2013 that establishes objectives for using small businesses tailored to spending patterns in each organization. The methodology for setting Performance Expectations objectives seeks to identify on an annual basis the industries where the Air Force has the most potential to increase small-business use, with particular attention paid to contracts with other-than-small businesses (OTSBs) that could be shifted to small businesses. This is part of a broader portfolio of initiatives that the Air Force Office of Small Business Programs is developing to help the Air Force change the way it approaches small-business contracting to develop greater competition for Air Force purchases.[8]

The Air Force Office of Small Business Programs asked RAND Project AIR FORCE to review the new method for setting Performance Expectations objectives and to recommend

---

[6] "Small" businesses are those below a size standard specified by the Small Business Administration (SBA) for their industry. Typically, the SBA defines small businesses as those with fewer than 500 employees (many manufacturing industries) or with less than $7.5 million in annual receipts (many service industries). Size standards do vary by specific industry, as delineated by six-digit North American Industry Classification System (NAICS) codes. For size standards by NAICS code, see U.S. Small Business Administration, 2014.

[7] In 2013, the small-business goal for DoD was 22.5 percent. That goal was divided between the various services and defense agencies based on how much and what they buy (e.g., large weapons, such as fighter airplanes, aircraft carriers, submarines, and tanks versus food, clothing, small industrial items, and facility support services) and how amenable their purchases may be to small-business use. The Army goal was 26.5 percent; that for the Navy was 16.5 percent, and that for the Air Force was 15.0 percent (U.S. Department of Defense Office of Small Business Programs undated c; Stackley, 2012; McDade, 2013).

[8] Related unpublished research that we conducted with colleagues Lloyd Dixon and Aaron Kofner found that small-business use is highly correlated with the distribution of appropriations to major budget categories. In particular, it found that small-business use tended to increase with Operations and Maintenance (O&M) but decrease when Procurement increased. Lower use of small businesses in procurement spending historically was a primary motivation for the new Performance Expectations methodology.

possible improvements, paying special attention to small-business purchases by industry as delineated by six-digit NAICS codes, Product or Service Codes (PSCs), budget categories, and industry trends. The purpose of this analysis is to increase the methodology's accuracy, and the analysis and the ensuing recommendations are not intended to increase or decrease the goals themselves. This report provides details of our assessment and recommendations.

Organization of This Report

In Chapter Two, we examine the Air Force Performance Expectations methodology and its goals. In Chapter Three, we review Air Force expenditures with small businesses and OTSBs by MAJCOMs and PEOs and by industry and PSC. We explore areas where more refined categories may be needed and conclude with recommended refinements to the methodology.

2. Current Performance Expectations Methodology

When setting small-business goals, federal authorities, and therefore the Air Force, use goaling dollars rather than all contract spending. The definition of *goaling dollars* excludes a number of spending categories.[9] All spending data in this report is limited to goaling dollars.

The Air Force's current Performance Expectations methodology breaks out spending by markets, which are combinations of industries designated through NAICS and goods designated by PSCs. Both NAICS codes and PSCs are entered on individual contract-action reports.[10] Each NAICS/PSC combination specifies an individual market. The combinations enable greater precision in the definition of buying markets than either metric would yield by itself.[11]

The Air Force Performance Expectations methodology is an eight-step process. It begins with establishing past performance in small-business purchasing and then seeking improvements in targeted areas. Table 2.1 outlines the eight steps of the process, each of which we discuss below.

---

[9] The SBA excludes some types of procurements from the percentage of small-business use in its goaling program. (Federal Procurement Data System—Next Generation, 2014). Exclusions include

- contracts performed outside of the United States and its territories
- acquisitions by agencies on behalf of foreign governments or entities or international organizations, such as Foreign Military Sales
- contracts funded primarily with agency-generated funds—that is, nonappropriated funds from such operations as user fees, gifts, etc. (this category includes TRICARE, the Defense Commissary Agency, the U.S. Postal Service, the Federal Aviation Administration, and the Transportation Security Administration)
- internal transactions between agencies that are not contracts
- mandatory sources specified in statutes, such as Federal Prison Industries and the Javits-Wagner-O'Day Program.

[10] NAICS codes indicate the industry within which the business is providing goods and services. PSCs indicate the type of good or service purchased. PSCs are more finely grained indicators of goods and services than NAICS codes.

[11] Both NAICS codes and PSCs are fairly broad categories in the economic industries for the good or service that they define. Industries defined by NAICS codes can vary widely in size, manufacturing processes, customer type, and other characteristics. For example, the aircraft-manufacturing industry (NAICS code 336411) includes small general-aviation aircraft, large commercial aircraft, military aircraft, and blimps, as well as maintenance and repair. In total, there were 1,065 NAICS codes in 2012 and 2,896 PSCs in October 2011. Combining the two provides more granularity that either one by itself, enabling better targeting of prospective markets to improve small-business use.

Table 2.1. Steps in Current Performance Expectations Methodology

1. Tabulate past fiscal year (FY) DoD and Air Force goaling-dollar spending and small-business use.

2. Distinguish Accessible markets (NAICS/PCS combinations where the Air Force has at least 1 percent small-business participation).

3. Distinguish Inaccessible markets (NAICS/PSC combinations where the Air Force has less than 1 percent small-business participation).

4. Identify High-Opportunity Target (HOT) Spots (NAICS/PSC combinations where DoD small-business use, excluding the Air Force, is more than double Air Force small-business use). HOT Spots can be in Accessible or Inaccessible markets.

5. Identify expiring OTSB Air Force contracts.

6. Identify expiring OTSB HOT Spot and Accessible minus HOT Spot contract dollars.

7. For each MAJCOM and PEO:
   a. Identify past-year small-business use in dollars.
   b. Add 10 percent of expiring OTSB Accessible-market HOT Spot dollars.
   c. Add 1 percent of expiring OTSB Accessible-market not–HOT Spot dollars.

8. Set MAJCOM/PEO performance expectation to total dollars from Step 7 divided by total past-year obligations.

The first step of the Performance Expectations methodology tabulates total DoD and Air Force spending as reported to the Federal Procurement Data System–Next Generation (FPDS-NG) on prior-year Air Force contract obligations. Because the methodology is focused on annual incremental increases in small-business use, it builds on the previous year of contract obligations only. The methodology uses the FPDS-NG Small Business Achievements by Awarding Organization Report data, which contain all DoD goaling dollars, including those of the Air Force. Because a very large volume of contracts are awarded and submitted to FPDS-NG in the last months of the FY, and because it takes six weeks or more for the Air Force to certify the FPDS-NG data as accurate and complete, the methodology is not used to set the current FY Performance Expectations until December or January—two to three months after the new FY starts.

As noted above, the Performance Expectations methodology tabulates expenditures by NAICS/PSC combinations. Each combination consists of one NAICS and one PSC. They are derived from contract actions, so if a contract includes three different contract actions, each with three different PSCs in the same NAICS, that contract would yield three NAICS/PSC combinations for the subset of contract dollars on each action. The FPDS indicates 8,973 NAICS/PSC combinations for purchases of Air Force goods and services in FY 2013.

The Performance Expectations methodology requires calculating Air Force small-business use within each NAICS/PSC combination. This calculation is necessary for the second and third steps of the methodology, distinguishing between Accessible and Inaccessible markets. Accessible markets are defined as those NAICS/PSC combinations where the Air Force spends at least 1 percent of its goaling dollars with small businesses. Inaccessible markets are defined as those where the Air Force spends less than 1 percent of its goaling dollars with small businesses.

The fourth step is to identify a list of HOT Spots from the list of Accessible and Inaccessible markets. HOT Spots are those NAICS/PSC combinations where DoD small-business use (excluding the Air Force) is *more than double* Air Force small-business use. These HOT Spots are intended to be NAICS/PSC combinations where the Air Force might have significant opportunities to increase its small-business use.

The fifth step is to identify expiring OTSB contracts at each MAJCOM and PEO in the FY for which the Air Force is developing a Performance Expectation objective.[12] The Air Force focuses on expiring OTSB contracts because they offer opportunities to increase small-business use in the current FY. Contracts still in force limit the ability of the Air Force to significantly change small-business use in any one year.[13]

The sixth step is to identify which expiring OTSB contracts are in markets deemed Accessible HOT Spots, as well as all other Accessible expiring contracts, including small-business and OTSB contracts.[14]

In the seventh step, the Air Force identifies (a) what each MAJCOM or PEO spent on small businesses in the past year, (b) 10 percent of the dollars that are in all of its expiring OTSB contracts for Accessible HOT Spot markets, and (c) 1 percent of the dollars that are in all of its expiring OTSB contracts for Accessible not–HOT Spot markets.

In the eighth step, the Air Force sums these three categories of dollars and divides them by all goaling dollars spent in the past year to determine the Performance Expectation objective for the MAJCOM or PEO in the subsequent year.

Figure 2.1 illustrates the methodology for identifying the portion of Air Force spending on which the Performance Expectations methodology focuses. It restricts itself to goaling dollars, eliminating obligations that do not count toward meeting Air Force small-business goals. It classifies each NAICS/PSC combination (or market) as Inaccessible or Accessible and identifies HOT Spots for the Air Force within Accessible markets. Within goaling dollars for each NAICS/PSC combination (or market), it further focuses on OTSB contracts that expire in the next FY.

---

[12] Because of delayed input of contract actions for the previous year, as discussed above, the methodology is actually identifying contracts expiring in the current year for which the Performance Expectation is being calculated.

[13] In FY 2013, more than 78 percent of Air Force obligations were made on contracts written before FY 2013, for which small-business use was only 12 percent. By contrast, for obligations made on contracts written in FY 2013, Air Force small-business use was 22 percent. The higher small-business use in contracts awarded in FY 2013 is typical of new contracts and actually lower than the small-business use in new FY 2012 contracts, which was 31 percent.

[14] While the Performance Expectations methodology does not include Inaccessible markets in the calculation, the Air Force Office of Small Business Programs finds value in this distinction to focus MAJCOMs and PEOs on the markets with the most opportunity to increase small-business use.

Figure 2.1. Current Performance Expectations Methodology

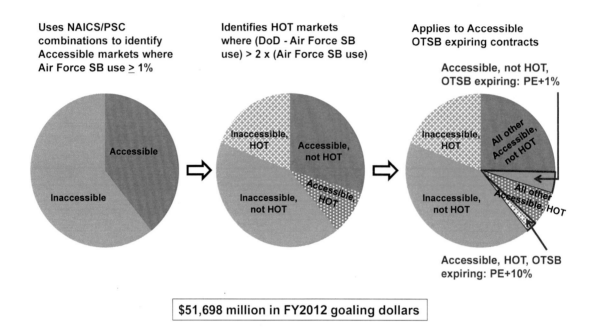

$51,698 million in FY2012 goaling dollars

NOTES: PE = Performance Expectation, SB = small business.

Table 2.2 provides a notional example of how the Performance Expectations methodology might apply to a hypothetical MAJCOM or PEO (which we refer to as a MAJCOM/PEO) that had spent $1 billion (goaling) in the past year, evenly divided among eight NAICS/PSC combinations (or $125 million each). The "Total" row illustrates the small-business spending for the MAJCOM/PEO as a whole, while the rows labeled "1" through "8" refer to each of the eight NAICS/PSC combinations in which the MAJCOM/PEO obligates money. The hypothetical MAJCOM/PEO had $277.5 million in overall small-business spending during the past year, accounting for 27.8 percent of its goaling spending (Step 1).[15]

---

[15] We later discuss details of small-business spending by MAJCOM. Here we note that MAJCOM spending with small businesses in FY 2013 ranged from 6.2 percent in the Air Force Space Command to 83.3 percent in the Air Force Reserve Command.

Table 2.2. Example of Performance Expectations Methodology Categorization and Calculations for a Notional MAJCOM/PEO

NAICS/PSC	% of Air Force Small-Business Use	Accessibility	% of DoD Small-Business Use	HOT Spot?[a]	Small-Business Spending (in millions)	Expiring OTSB Contracts (in millions)	Multiple for Performance Expectations Calculation	Resulting Dollars for Performance Expectations Calculation (in millions)	New Small-Business Use % with Performance Expectations Calculation
1	100.0	Accessible	100.0	No	125.00	0.0	0.01	125.00	100.0
2	90.0	Accessible	90.0	No	112.50	12.5	0.01	112.60	90.1
3	20.0	Accessible	60.0	Yes	25.00	25.0	0.10	27.50	22.0
4	10.0	Accessible	30.0	Yes	12.50	25.0	0.10	15.00	12.0
5	0.9	Inaccessible	0.9	No	1.13	25.0	0.00	1.10	0.9
6	0.8	Inaccessible	0.9	No	1.00	25.0	0.00	1.00	0.8
7	0.2	Inaccessible	0.6	Yes	0.25	25.0	0.00	0.25	0.2
8	0.1	Inaccessible	0.3	Yes	0.13	25.0	0.00	0.13	0.1
Total	27.8[b]				277.50	162.5		282.50	28.3[b]

[a] All calculations for the HOT Spot determination are not visible in this notional table. HOT Spots are those NAICS/PSC combinations where DoD small-business use (excluding the Air Force) is more than double Air Force small-business use.

[b] This total represents the percentage of all, not the sum. This table assumes that MAJCOM/PEO spent $1 billion in goaling dollars in previous year, divided evenly ($125 million each) among the eight listed NAICS/PSC combinations.

The second column shows the small-business use rate for the total Air Force in each of the NAICS/PSC combinations. Small-business use in these eight combinations varied from 0.1 percent to 100 percent. In column three, the top four hypothetical markets shown are deemed Accessible—i.e., they have Air Force small-business expenditures of at least 1 percent of goaling dollars (Step 2)—and four are deemed Inaccessible—i.e., they have Air Force small-business expenditures of less than 1 percent of goaling dollars (Step 3).

The fourth column shows the DoD small-business use rate, excluding the Air Force, for the NAICS/PSC combination in each row. This is used to calculate which NAICS/PSC markets are HOT Spots (Step 4). Column five shows that four of the markets—two of which are Accessible and two Inaccessible—are deemed HOT Spots, where the DoD small-business use rate (excluding the Air Force) is at least twice the Air Force rate.

We next assume that each NAICS/PSC combination has $25.0 million in expiring OTSB contracts (columns six and seven), except for the first two NAICS/PSC combinations listed (Step 5). The first NAICS/PSC combination, which has 100-percent small-business use, has $0 in expiring OTSB contracts. The second NAICS/PSC combination, which has 90-percent small-business use, has $12.5 million in expiring OTSB contracts.

To calculate the Performance Expectations for our hypothetical MAJCOM/PEO, we sum for each NAICS/PSC combination:

- the value of expiring OTSB contracts (column seven), multiplied by a corresponding multiple for the Performance Expectations calculation (column eight)
  - 0.00 for Inaccessible NAICS/PSC combinations
  - 0.01 for Accessible combinations that are not HOT Spots
  - 0.10 for Accessible combinations that are HOT Spots (Steps 6 and 7)
- small-business spending in the past year (in millions of dollars, shown in column six).

Column nine shows the results of this summation. For all of the NAICS/PSC markets, the total is $282.5 million. This sum of $282.5 million is then divided by the $1 billion in total goaling spending in the previous year that we noted for this hypothetical MAJCOM/PEO. Shown in column ten, this generates the percentage of small-business use that the Performance Expectations methodology would expect from the hypothetical MAJCOM/PEO in the coming year (Step 8). The Performance Expectations methodology indicates that our hypothetical MAJCOM/PEO should aim to contract at least 28.3 percent of its obligations with small businesses.

We analyzed the current methodology to identify where and how the Performance Expectations methodology might be improved and refined. First, we developed a baseline of recent small-business use by MAJCOM and PEO. This includes analysis of total and small-business purchases by MAJCOM and PEO in FY 2012 and FY 2013. Next, we reviewed the current Performance Expectations methodology and identified possible areas where it might be improved, many of which focus on identifying relevant markets for analysis. Third, we identified

8

available data for application of a revised methodology and reviewed their benefits and drawbacks. Finally, we incorporated our data analyses into an improved methodology and tested it by making and comparing Performance Expectations for FY 2013. Our goal in each of these steps was to identify areas where the methodology might be made more accurate.

3. Air Force Small-Business Spending by MAJCOM and PEO

The Air Force Office of Small Business Programs (SAF/SB) tracks Air Force contracts that are written, used, and obligated by MAJCOMs and other Air Force organizations.[16] Organizations are listed in Table 3.1 in order of their share of Air Force purchases (from high to low). The table also lists goaling dollars spent in all purchases and with small businesses for FY 2013.

Table 3.1. Total Air Force Goaling Dollars Spent by MAJCOM, Overall and with Small Businesses, FY 2013

Organization	Total Goaling Dollars Spent	Total Small-Business Dollars Spent
Air Force Materiel Command (AFMC)	32,163,622,533	3,481,325,208
Air Force Space Command (AFSPC)	7,199,483,354	448,375,959
Air Combat Command (ACC)	1,231,410,756	673,583,607
Air Education and Training Command (AETC)	998,148,244	398,691,420
Air Mobility Command (AMC)	712,461,025	504,601,787
Pacific Air Forces (PACAF)	321,211,842	143,751,105
Air Force District of Washington (AFDW)	281,873,460	82,581,615
Air Force Intelligence, Surveillance, and Reconnaissance Agency (AFISRA)	245,551,947	70,886,304
Air Force Global Strike Command (AFGSC)	193,698,013	146,853,563
United States Air Force Academy (USAFA)	190,332,871	135,547,829
Air Force Reserve Command (AFRC)	182,001,592	151,556,339
Air Force Special Operations Command (AFSOC)	143,860,598	94,480,134
United States Air Forces in Europe (USAFE)	26,285,574	9,276,888
Air Force Operational Test and Evaluation Center (AFOTEC)	7,769,133	2,242,029
TOTAL Air Force goaling dollars	43,897,710,942	6,343,753,788

SOURCE: FY 2013 FPDS-NG DoD goaling dollars.

---

[16] We use the term *MAJCOMs* to refer to major commands and other Air Force organizations for which SAF/SB provides annual small-business Performance Expectation goals. Requirements from external DoD organizations that transfer funds to Air Force MAJCOMs are included in Air Force small-business use rates. Conversely, contracts written by external DoD organizations that are funded by the Air Force are not included in Air Force small-business use rates per SBA goaling dollar inclusion and exclusion rules.

Figure 3.1 shows the percentages that each MAJCOM had of all Air Force goaling dollar obligations (green columns keyed to left axis) and of all Air Force small-business obligations (red columns keyed to left axis), as well as levels of small-business use in each (dark blue squares keyed to right axis) for FY 2013. Each MAJCOM is arranged from left to right in order of its percentage of Air Force purchases.

Figure 3.1. Performance Expectations Methodology Needs to Account for Wide Variation Across MAJCOMs

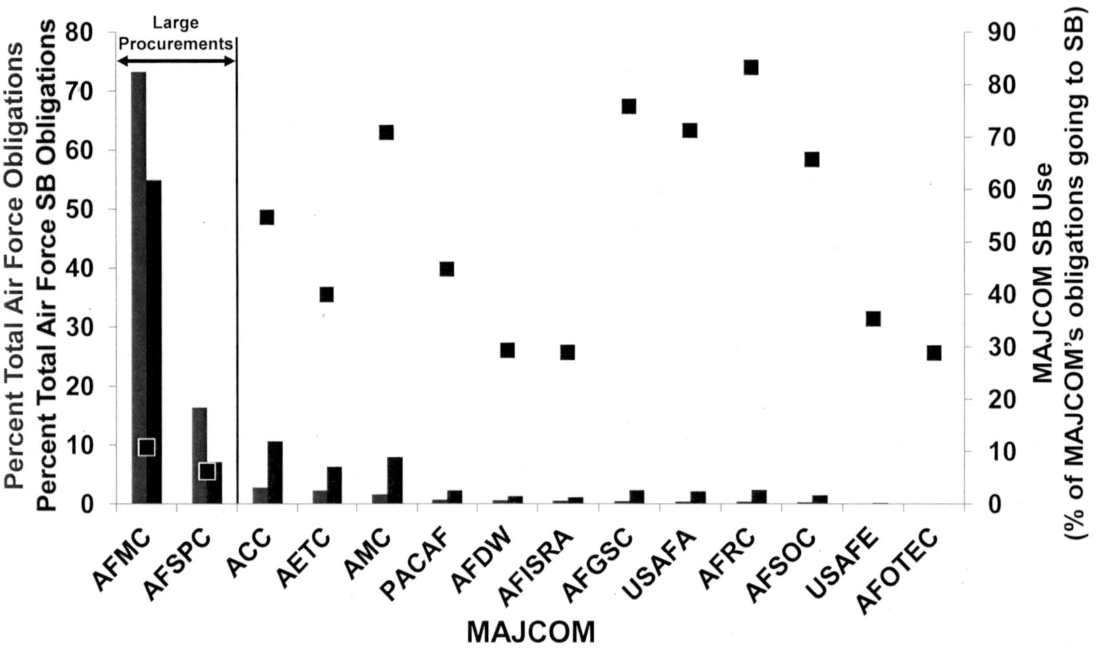

SOURCE: FY 2013 FPDS-NG DoD goaling dollars as of January 2014.
NOTE: Percentages shown are of goaling dollars.

Generally, small-business use has been higher in MAJCOMs that account for smaller percentages of Air Force dollars. In fact, in all MAJCOMs but AFMC and AFSPC, small-business use has been above the federal goal of 23 percent. Small-business use within AFMC and AFSPC has likely been low because they obligate more dollars for Procurement and for Research, Development, Test, and Evaluation (RDT&E) than any other command.[17] These obligations reflect AFMC's and AFSPC's use of OTSB prime contractors in recent years.

Part of the motivation for the Performance Expectations methodology is the desire to break the patterns of past trends by identifying pockets of opportunity to increase small-business spending in areas that have traditionally seen lower usage.

---

[17] Small-business use in these budget categories is traditionally lower than in other budget categories (Moore, Grammich, and Mele, 2014, p. 24). We address some of the reasons behind this, including large weapon systems without breakout of smaller components, later in the report.

Over time, the percentage distribution of overall spending and small-business goaling dollar spending by the Air Force is relatively consistent within most organizations. Table 3.2 shows three groups of percentages: total Air Force spending, Air Force spending obligated to small businesses, and MAJCOM spending obligated to small businesses. Each percentage is shown for 2012 and 2013. Two commands—AFMC and AFSPC—account for nearly 90 percent of all Air Force (goaling-dollar) procurement, as well as roughly 60 percent of small-business procurement. Thus, these two commands obligate many more dollars to small business than other commands, even though their small-business use rate is much lower. AFMC and AFSPC therefore have a much larger effect on total Air Force small business use.

Table 3.2. Little Change from FY 2012 to FY 2013 in Small-Business Performance Across MAJCOMs

MAJCOM	% of Air Force Spending		% of Air Force Small-Business Spending		% of MAJCOM Contract-Action Spending with Small Businesses	
	FY 2012	FY 2013	FY 2012	FY 2013	FY 2012	FY 2013
AFMC	71.1	73.3	51.1	54.9	10.6	10.8
AFSPC	17.9	16.4	7.6	7.1	6.3	6.2
ACC	2.9	2.8	11.3	10.6	57.2	54.7
AETC	2.5	2.3	8.7	6.3	51.2	39.9
AMC	1.7	1.6	7.8	8.0	68.1	70.8
PACAF	0.8	0.7	2.6	2.3	49.5	44.8
AFDW	0.8	0.6	1.4	1.3	26.0	29.3
AFISRA	0.5	0.6	0.9	1.1	25.3	28.9
AFGSC	0.4	0.4	2.3	2.3	75.3	75.8
USAFA	0.3	0.4	1.6	2.1	75.1	71.2
AFRC	0.5	0.4	2.3	2.4	69.4	83.3
AFSOC	0.5	0.3	2.2	1.5	70.9	65.7
USAFE	0.1	0.1	0.1	0.1	19.6	35.3
AFOTEC	<0.1	<0.1	0.1	<0.1	59.6	28.9

SOURCE: FY 2012–FY 2013 FPDS-NG DoD goaling dollars.

The percentage distribution of total Air Force spending and of Air Force small-business spending across the MAJCOMs changed little from FY 2012 to FY 2013, despite budget turmoil from continuing resolutions and sequestration. Observed minor changes in the distribution of small-business spending may be due to the types of contracts that expired in FY 2013. Decreases may be a result of greater emphasis on mission-essential spending, particularly during budget continuations and sequestration, as DoD and Air Force leaders have emphasized weapon-system procurement in recent years (see, for example, Carter, 2013; Wright, 2012; and Gates, 2011). More than 65 percent of Air Force goaling dollars are spent by PEOs, which report to the assistant secretary of the Air Force (Acquisition) but use contracting offices within AFMC and AFSPC with responsibility for a specific program or a portfolio of similar programs.[18] In order of share of Air Force purchases in FY 2013, Air Force PEOs, as defined by SAF/SB in its 2013 and 2014 Performance Expectations objectives, are shown in Table 3.3.

---

[18] The PEOs in this list match those used by SAF/SB to calculate its 2013 and 2014 Performance Expectations objectives. We maintain its use here to compare the effects of our recommended changes to the methodology with the existing methodology.

Table 3.3. Total Air Force Goaling Dollars Spent by PEO, Overall and with Small Businesses, FY 2013

PEO	Total Goaling Dollars Spent	Total Small-Business Dollars Spent
Space Systems	5,052,809,313	150,815,687
Mobility	5,004,794,659	32,586,953
Intelligence, surveillance, and reconnaissance and Special Operations Forces (ISR and SOF)	4,998,044,799	249,045,527
Fighter Bomber	3,501,770,780	29,834,611
Battle Management	2,047,532,578	129,104,652
Tankers	1,764,133,108	6,878,766
Agile Combat Systems	1,409,611,896	175,812,761
Space Launch	1,371,958,396	522,669
Weapons	1,214,064,854	29,305,555
Command, control, communications, intelligence, and networks (C3I&N) strategic systems	778,515,739	85,570,180
Strategic Systems	567,607,581	8,996,416
Space	372,916,174	11,768,185
Business and Enterprise Systems (B&ES)	348,120,757	113,207,096
Combat and Mission (CM) Support	218,554,490	113,932,479
Other Air Force	15,247,275,820	5,206,372,250
TOTAL Air Force goaling dollars	43,897,710,942	6,343,753,788

SOURCE: FY 2013 FPDS-NG DoD goaling dollars.
NOTES: Space PEO expenditures are for two contracting offices at the Air Force Life Cycle Management Center within the Air Force Materiel Command that appear in FY 2013 data but not in FY 2012 data. Space Systems and Space Launch PEO expenditures are by two separate groups of contracting offices in the Space and Missile Center of the Air Force Space Command. Note that PEO/CM does not have any obligation authority (the FY 2013 data does show a few contracts attributed to PEO/CM, but these contracts were actually under another PEO). All other PEOs have a dedicated contracting office/DoD Activity Address Code; PEO/CM cuts across the Air Force enterprise and across many contracting offices.

Figure 3.2 shows the percentage each PEO had of all Air Force goaling expenditures (green columns keyed to the left axis) and the percentage each PEO had of all Air Force small-business expenditures (red columns keyed to left axis). The figure also shows the levels of small-business use in each PEO as the percentage of PEO dollars obligated to small businesses (dark blue squares keyed to the right axis) in FY 2013. Given that PEOs do not account for all Air Force expenditures, the percentages of Air Force expenditures and of Air Force small-business expenditures will not equal 100 percent.

Figure 3.2. Nearly All PEOs Have Low Small-Business Use Rates

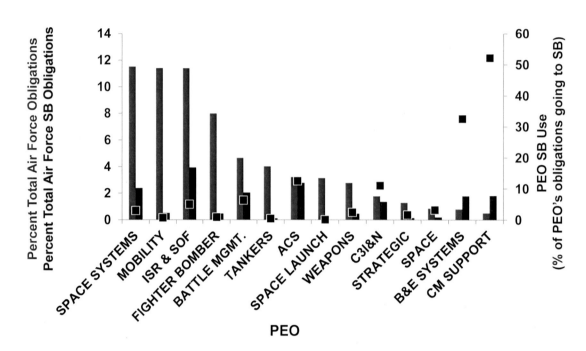

SOURCE: FY 2013 FPDS-NG DoD goaling dollars as of January 2014.
NOTE: Percentages shown are of goaling dollars, and Space PEO expenditures are those for two contracting officers at the Air Force Life Cycle Management Center within the Air Force Materiel Command that appear in FY 2013 data but not in FY 2012 data.

Because PEOs often manage large weapon systems, it is not surprising that their levels of small-business use have been low. In FY 2013, small-business use rates were below 1 percent for four PEOs and below 5 percent for nine PEOs. Nevertheless, two PEOs—albeit the two with the smallest shares of goaling dollars among Air Force PEOs—had small-business use rates exceeding 30 percent. The CM Support PEO purchased most of its goods and services through small businesses.

Spending by PEO was relatively concentrated, as shown by the PEO percentage of total Air Force obligations (green columns). Three PEOs—Space Systems, Mobility, and ISR and SOF— accounted for more than one in three goaling dollars spent by the Air Force in FY 2013. They also accounted for the majority of goaling dollars for all PEOs.

While PEOs accounted for nearly two-thirds of all Air Force goaling dollars, they accounted for only about one-sixth of dollars that the Air Force spent with small businesses (red columns). Space Systems, Mobility, and ISR and SOF were responsible for only about one-third of small-business dollars spent by the PEOs.

Recent initiatives within SAF/SB, including the new Performance Expectations methodology, seek to find additional short- and long-term opportunities for small businesses to compete for contracts within PEO portfolios. Particular emphasis falls on rethinking how

16

requirements are specified or packaged to break out portions of work that are appropriate for competition among small businesses.

Similarly, Table 3.4 shows that the percentage distribution of total Air Force goaling dollar spending by PEO is relatively steady over time. The table presents three groups of percentages: the percentage of total Air Force spending, the percentage of Air Force spending obligated to small businesses, and the percentage of PEO spending obligated to small businesses. The table includes percentages for both 2012 and 2013. For most PEOs, the percentage share of Air Force goaling dollars changed little from FY 2012 to FY 2013, despite budget turmoil caused by continuing resolutions and sequestration.

Table 3.4. PEOs Have Relatively Few Small-Business Dollars

PEO	% of Air Force $		% of Air Force Small-Business $		% of PEO Contract-Action $ with Small Businesses	
	FY 2012	FY 2013	FY 2012	FY 2013	FY 2012	FY 2013
Space Systems	10.8	11.5	2.5	2.4	3.4	2.9
Mobility	12.1	11.4	0.7	0.5	0.8	0.7
ISR and SOF	11.0	11.4	1.8	3.9	2.5	5.0
Fighter Bomber	6.4	8.0	0.4	0.5	1.0	0.9
Battle Management	5.3	4.7	1.3	2.0	3.8	6.3
Tankers	1.6	4.0	<0.1	0.1	0.1	0.4
Agile Combat Systems	3.3	3.2	2.4	2.8	10.5	12.5
Space Launch	5.1	3.1	<0.1	<0.1	<0.1	<0.1
Weapons	2.8	2.8	0.8	0.5	4.3	2.4
C3I&N	2.1	1.8	0.8	1.3	5.7	11.0
Strategic Systems	1.4	1.3	0.1	0.1	1.1	1.6
Space	n/a	0.8	n/a	0.2	n/a	3.2
B&E Systems	1.3	0.8	2.0	1.8	22.3	32.5
CM Support	2.4	0.5	1.5	1.8	9.2	52.1

SOURCE: FY 2012–FY 2013 FPDS-NG DoD goaling dollars.

These analyses of spending demonstrate the need for separate small-business goals for these different components of the Air Force. The current Performance Expectations methodology addresses this variation effectively, and such accounting needs to be maintained through any refinements to the methodology. We turn next to examine other factors that could be accounted for as well.

4. Air Force Small-Business Spending by Budget Category and Industry

Small-business use differs across the Air Force's budget categories. Further, the Air Force has a different distribution of spending by budget category than other parts of DoD, and its small-business use differs from that elsewhere in DoD. This raises questions about using DoD small-business spending as a whole as a standard for identifying HOT Spots for the Performance Expectations methodology.

Table 4.1 shows the amount of spending in each budget category and the small-business use rate in each budget category for DoD as a whole, the three military services, other defense agencies (ODA), and the DLA. (The column for DoD as a whole includes the three services, ODA, and DLA.) For example, for DoD as a whole, 48.6 percent of spending was in the O&M budget category, and 28.8 percent of this was spent on small businesses. In FY 2013, the three services had the highest percentage of small-business use in the categories of Family Housing and O&M. Within the Air Force and Navy, small-business use was greater in O&M than in any other category but Family Housing. Most spending for Family Housing in the Air Force, Army, and Navy is with small businesses, but Family Housing is a very small proportion of the DoD budget.

Across DoD, DLA had the highest percentage of small-business use overall. Almost all of its budget was in O&M, which affords many small-business opportunities. DLA buys more goods amenable to small-business production, such as food; clothing; small hardware (such as nuts, bolts, and screws); and consumable weapon system items, such as filters and hoses. Indeed, DLA was created specifically to consolidate the buying of common or similar goods across DoD.[19] The services are supposed to purchase these types of goods through DLA unless they need to procure something highly specialized or an item unique to a service, such as body armor.

---

[19] DoD and the services are also attempting to consolidate the buying of similar goods and services via strategic sourcing, but the buying categories that they are trying to consolidate, particularly those such as professional services, are typically different from the goods that DLA is responsible for buying.

Table 4.1. Small-Business Use by Budget Category Varies Across DoD

% FY 2013

Budget Category	DoD Spend	DoD SB Use	Air Force Spend	Air Force SB Use	Navy Spend	Navy SB Use	Army Spend	Army SB Use	ODA Spend	ODA SB Use	DLA Spend	DLA SB Use
O&M[a]	48.6	28.8	44.8	22.7	35.5	30.6	42.9	33.0	67.6	16.6	99.4	37.8
Procurement	29.8	6.0	26.0	3.9	49.3	4.0	28.7	10.2	5.8	17.5	0.4	17.3
RDT&E	15.0	14.9	28.5	11.0	12.4	12.1	10.8	28.9	20.7	11.1	0.1	82.5
Military Construction	3.0	41.2	0.4	13.2	2.3	28.8	7.8	46.8	0.2	20.8	0.0	0.0
Family Housing	0.1	47.6	0.1	56.6	0.1	53.2	0.0	59.7	0.1	0.0	0	0
Non-DoD (Office of the Corps of Engineers [OCE] and other)	2.6	34.8	0.2	0.5	0.1	23.4	9.6	35.1	0.1	5.2	0.0	−50[b]
All[c]	100	**20.2**[d]	100	**14.5**	100	**15.1**	100	**27.4**	100	**14.6**	100	**37.8**

SOURCES: Air Force, Army, and Navy FPDS-NG data underlying standard report "Small Business Achievements by Awarding Organization"; DoD ODA, DLA from FY 2013 FPDS-NG DoD goaling dollars as of January 2014.

NOTE: Bold text indicates overall small-business goal attainment.

[a] Includes trust funds.

[b] This is a result of deobligations exceeding obligations in the modest small-business expenditures for this category.

[c] Includes non-DoD OCE and other funds spending.

[d] DoD goal in FY 2013 was 22.5 percent.

The Army had the next-highest small-business use overall. It had a higher proportion of its budget in Military Construction and non-DoD purchases, including purchases for the OCE, both of which have relatively high rates of small-business use.

Overall levels of small-business use in the Air Force, Navy, and ODA were fairly similar but still varied by budget category. The percentage of expenditures with small businesses for these agencies also varied within budget category.

These variations indicate that DoD may not be the best comparison group for identifying HOT Spots where the Air Force could significantly increase its small-business use. Eliminating goods that the services do not typically buy, such as goods purchased by DLA, from the comparison should make the comparisons more realistic.

The data above also point to the need to account for budget categories when setting small-business performance expectations because of the varying levels of amenability of goods and services to small-business procurement.

The Air Force's use of small business also varies within industry by budget category. While 460 of the 780 industries (as defined by six-digit NAICS codes) in which the Air Force purchased goods and services were associated with only one budget category, 320 were associated with two or more budget categories. Accounting for budget category thus offers a way to further refine industry groupings and better identify targets for increased small-business usage.

Table 4.2 lists the top 12 industries by spending of goaling dollars for the Air Force in FY 2013.[20] In each of these industries, the Air Force had expenditures in multiple budget categories. The table shows the total goaling dollars obligated in each industry and the percentage of goaling dollars spent with small businesses in each industry highlighted in yellow. It also shows, for the largest three Air Force budget categories, the percentage of total obligations in each budget category and the percentage of those dollars spent with small businesses.

---

[20] For this analysis, we treat NAICS codes 541710, 541711, and 541712 as one industry. The Census Bureau used NAICS code 541710 through the 2002 Economic Census for businesses in Research and Development in the Physical, Engineering, and Life Sciences. For the 2007 Economic Census, it divided businesses in NAICS code 541710 into two mutually exclusive and exhaustive (in reference to NAICS code 541710) industries, both with new NAICS codes: Research and Development in Biotechnology (NAICS code 541711) and Research and Development in the Physical, Engineering, and Life Sciences (except Biotechnology) (NAICS code 541712). However, years later, contract-action reports continue to use NAICS code 541710, possibly because such reports are for actions on contracts awarded when 541710 was still an active NAICS code. For example, in FY 2013, Air Force contract-action reports indicate $2.013 billion for firms with NAICS code 541710, which no longer exists, in addition to $4.394 billion for firms with NAICS code 541712 and $45 million for firms with NAICS code 541711.

Table 4.2. Air Force Small-Business Use Varies by Budget Category Within a Single Industry

Industry Name (industries defined by six-digit NAICS codes)	Total		Procurement		RDT&E		Operations and Maintenance	
	Goaling in Millions of $	% SB	% of $	% SB	% of $	% SB	% of $	% SB
Aircraft Manufacturing	8,916	2.3	40.6	0.6	27.8	2.5	30.8	4.6
Research and Development in the Physical, Engineering, and Life Sciences[a]	6,452	19.8	6.4	27.7	77.8	20.0	15.6	15.1
Other Aircraft Parts and Auxiliary Equipment Manufacturing	4,208	3.9	29.4	1.7	5.1	1.2	64.1	5.2
Engineering Services	3,305	8.1	15.2	6.9	28.2	6.8	56.5	9.0
Guided Missile and Space Vehicle Manufacturing	2,594	0.0	82.5	0.0	13.0	0.0	4.1	0.0
Other Support Activities for Air Transportation	2,212	1.3	11.0	0.2	3.0	6.8	81.0	1.2
Search, Detection, Navigation, Guidance, Aeronautical, and Nautical System and Instrument Manufacturing	1,502	2.1	44.9	2.1	25.4	2.2	29.7	1.9
Nonscheduled Chartered Freight Air Transportation	1,285	0.0	92.9	0.0	7.1	0.0	0.0	n/a
All Other Professional, Scientific, and Technical Services	1,265	3.8	0.2	41.1	93.2	0.4	6.5	51.4
Facilities Support Services	916	34.1	0.1	19.9	12.3	11.4	85.8	36.0
Computer Systems Design Services	839	16.9	7.4	40.9	11.7	13.5	80.9	15.2
Wired Telecommunications Carriers	771	33.6	27.8	26.1	7.2	6.3	64.9	39.9

SOURCE: FY 2013 FPDS-NG data.
NOTE: The budget category that has the largest percentage of obligations in each industry is listed in bold.
[a] As noted, for purposes of this analysis, we treat NAICS codes 541710, 541711, and 541712 as representing one industry.

22

Aircraft Manufacturing provides a good example of variation within industry. In FY 2013, the Air Force obligated more goaling dollars in Aircraft Manufacturing (NAICS code 336411) than in any other industry. While the overall small-business usage rate was 2.3 percent, the rate varied across budget categories from 0.2 percent (residual categories not shown in table) to 4.6 percent (O&M). The most notable difference here is between O&M and Procurement; the proportion of O&M aircraft-manufacturing obligations with small business is nearly eight times that of Procurement. Differentiating among budget categories within the industry would help SAF/SB better tailor small-business use improvement objectives among spending segments.[21] This would allow for more aggressive goals in those areas where small businesses have been shown to successfully compete and reevaluation of the structure of requirements in other areas to encourage greater small-business participation.

These findings—that considerable spending occurs outside the leading budget category for an industry and that small-business spending can differ between the leading and other categories in an industry—illustrate that some industries may be too broad for setting small-business use goals. They also indicate that Air Force small-business use will likely vary within an industry by the distribution of purchases across budget categories. The differences we saw earlier between budget categories apply here as well. Most industries for which O&M is the leading spending category have higher overall levels of small-business use but lower levels of small-business use outside O&M.

As mentioned earlier, both PSCs and NAICS codes can cover a broad range of economic activity, and these codes may need refinement, particularly to analyze potential small-business opportunities. Defining markets with as much precision as possible enables small-business opportunities, and therefore small-business goals, to be as accurate as possible. We use two examples of what appear to be similar NAICS codes and PSCs to illustrate the breadth of some individual industries and PSCs. We calculate these from contract-action report forms that require contracting officers to list both the PSC and the NAICS code for the good or service purchased.

The NAICS code (541330) for the Engineering Services industry and the PSC (R425) for Professional Engineering/Technical Services appear to cover similar services. Yet the FY 2013 FPDS-NG shows more than 176 PSCs associated with the Engineering Services NAICS code and 53 NAICS codes associated with the PSC for Professional Engineering/Technical Services.

In another example, both a NAICS code and a PSC apply to facilities/building support services. Yet the FPDS-NG shows 89 different PSCs associated with the NAICS code (561210) for Facilities Support Services and 31 different NAICS codes associated with the PSC (S216) for Housekeeping—Facilities Operations Support.

---

[21] NAICS codes and definitions are reviewed and revised every five years prior to the Economic Census by the Economic Classification Policy Committee, which is staffed by the Bureau of Economic Analysis, the Bureau of Labor Statistics, and the Census Bureau. The DoD Office of Small Business Programs could clarify the challenges facing DoD use of small businesses by bringing these and other related findings to the Economic Classification Policy Committee's attention.

This may partially be the result of imprecise assignment of PSCs and NAICS codes to contract-action reports by contracting personnel. It may also be the result of some industry and PSC definitions that are too broad to adequately capture the variety of goods or services being purchased, as we will discuss in greater detail. Both have implications for efforts to set Performance Expectations.

Some of the PSC-NAICS code combinations also cross budget categories. Table 4.3 shows the number of budget categories per NAICS/PSC combination (column one), the percentage of all NAICS/PSC combinations that have that number of budget categories (column two), and the percentage of total Air Force spending and the percentage of spending devoted to small businesses for the NAICS/PSC combinations that have that number of budget categories (columns three and four). Among the markets defined by PSC-NAICS code combinations, 14 percent cross two or more budget categories, and a handful cross as many as seven categories.[22] Because small-business use can vary widely by budget category, this points to opportunities to further refine how markets are defined by splitting PSC-NAICS code combinations into budget categories as well. PSC-NAICS code markets that cross multiple budget categories accounted for 83 percent of Air Force goaling expenditures in FY 2013. Small-business expenditures within these markets varied from 1 to 29 percent. Such diversity among markets with multiple budget categories suggests that other opportunities to refine analysis for setting performance expectations may exist as well.

Table 4.3. Many NAICS/PSC Groups Are in One Budget Category, But 14 Percent of Groups Have High Spending and Multiple Categories

# of Budget Categories per NAICS/PSC Group	% of Air Force NAICS/PSC Groups	Obligations	
		% of Total Air Force	% of Small Business
7	0.0	0.6	3.8
6	0.0	0.2	1.2
5	0.2	7.3	8.9
4	0.7 (14%)	27.3 (83%)	10.5
3	3.2	27.6	29.0
2	10.0	20.5	22.4
1	82.7	16.5	24.2
0	3.0	0.0	0.0

SOURCE: FY 2013 FPDS-NG DoD goaling dollars as of January 2014.

---

[22] The NAICS/PSC combinations with no budget categories are those for which there may have been contract actions but no dollar obligations. Contract actions for such combinations typically do not receive Treasury Account codes, which are used to classify contract actions by budget category.

The current Performance Expectations methodology focuses on expiring OTSB contracts to identify potential targets of opportunity to increase small-business usage. Another refinement to the Air Force's approach is to consider market segments within OTSB contracts that already have extremely high levels of small-business participation. Table 4.4 shows small-business use rates by industry grouped into deciles (column one), the percentage of all NAICS in which the Air Force buys goods and services that have these small-business use rates (column two), the percentage of total Air Force spending in these groups of industries (column three), and the percentage of Air Force small-business spending in these groups of industries (column four). As Table 4.4 shows, in FY 2013, in half of the industries in which it purchased goods and services, the Air Force made at least 90 percent of its obligations with small businesses.[23] Because there is little room for the Air Force to improve small-business performance in these industries, they should not be included in calculating Performance Expectations. For many goods and services, current OTSB contractors may own the underlying intellectual property, preventing competition. The Air Force also may not want to drive industries to be 100 percent small business, and hence may wish to exclude industries approaching such a standard from its Performance Expectations calculations. Given constrained personnel resources, the Air Force should have its Performance Expectations methodology focus efforts on those areas where it is most likely to significantly improve its small-business use.[24]

---

[23] Air Force obligations with small businesses in an industry could exceed 100 percent. This can occur when deobligations to OTSBs exceed obligations to such businesses. Conversely, Air Force obligations with small businesses in an industry can be less than 0 percent when deobligations to small businesses exceed obligations to such businesses within it or when overall obligations in the industry are negative. We lump these anomalies together with their closest category in the table.

[24] Additional data on purchases by markets defined by combinations of NAICS codes and PSCs or by NAICS codes, PSCs, and budget categories show that many markets already have very high levels of small-business use but not very high obligations, while others have very low levels of such use but high levels of obligations.

Table 4.4. Small-Business Use Is at or Near Its Maximum in Many Industries

% of Small-Business Use in Industry	% of Air Force Industries (as defined by NAICS codes)	FY 2013 Obligations	
		% of Total Air Force	% of Small Business
100+	37.8 ⎫ 50%	1.2 ⎫ 3%	7.7 ⎫ 19%
90–99.9	12.1 ⎭	1.7 ⎭	11.4 ⎭
80–89.9	6.2	2.1	11.9
70–79.9	3.3	0.4	2.2
60–69.9	2.8	0.1	0.5
50–59.9	4.5	2.1	7.8
40–49.9	2.7	2.0	6.3
30–39.9	2.9	5.5	12.8
20–29.9	2.3	11.5	21.2
10–19.9	3.7	3.1	3.4
<10	15.2	70.4	14.8

SOURCE: FY 2013 FPDS-NG DoD goaling dollars as of January 2014.

Another indicator of the supply of competitive small businesses in industries in which the Air Force procures goods and services is the concentration of that industry—how much of the revenue generated is consolidated in a relatively small number of firms. If the business conducted in an industry is disproportionately concentrated in firms that are not small, then small businesses will likely have a harder time breaking in and competing.[25] Currently, many of the leading industries in which the Air Force purchases goods and services are highly concentrated. The degree of concentration or consolidation in an industry typically increases over time.[26] Thus, analyzing the degree of consolidation in an industry can help determine best strategies for purchasing in that industry. When industries are less concentrated, they are likely to respond more easily to efforts to increase small-business use. The Air Force may wish to develop alternative methods for increasing small-business competition in more-concentrated industries, such as breaking components out of large weapon system contracts.

Table 4.5 shows the 12 leading industries in which the Air Force purchased goods and services in FY 2013, listed in the order of their goaling dollars from the Air Force. These 12 industries accounted for 78 percent of all Air Force obligations in FY 2013 and 43 percent of the obligations to small businesses. We also show data from the 2007 Economic Census (the most

---

[25] For an understanding of the barriers to entry to a market, see the elements of industry structure in Porter (1985).

[26] For more on how many industries have consolidated over time, see Deans, Kroeger, and Zeisel (2002). For more on how to identify consolidated industries, see the appendix.

recent available at the time of publication) on the total number of firms within these industries, the number of OTSBs, and the percentage of OTSBs among all firms and among all receipts in the industry.[27] Note that the Economic Census measures a firm's receipt by its *primary* industry.

Table 4.5. OTSBs Dominate Many Industries in Which the Air Force
Makes Most of Its Purchases

Industry	Total Number of Firms	OTSBs		
		Number	% of Total Number	% of Total Industry Receipts
Aircraft Manufacturing	221	30	13.8	97.3
Research and Development in the Physical, Engineering, and Life Sciences	11,382	622	5.5	65.9
Other Aircraft Parts and Auxiliary Equipment Manufacturing	770	59	7.7	83.7
Engineering Services	47,714	4,886	10.2	84.4
Guided Missile and Space Vehicle Manufacturing	14	7	50.0	97.2
Other Support Activities for Air Transportation	2,986	274	9.2	82.2
Search, Detection, Navigation, Guidance, Aeronautical, and Nautical System and Instrument Manufacturing	494	53	10.7	94.5
Nonscheduled Chartered Freight Air Transportation	206	4	1.9	58.8
All Other Professional, Scientific, and Technical Services	15,547	203	1.3	36.2
Facilities Support Services	1,861	251	13.5	86.5
Computer Systems Design Services	37,028	622	1.7	75.4
Wired Telecommunications Carriers	3,471	77	2.2	86.1

SOURCE: Special tabulations of 2007 Economic Census for SBA.

---

[27] We calculate these statistics, as noted in the table, from special tabulations that the Census Bureau did for the SBA on firms and receipts by number-of-employee or size-of-receipts categories. We used the small-business size thresholds in place for these industries in 2007, the year of the data, ignoring exceptions to these thresholds that we discuss next. In some cases, the threshold fell within a category (e.g., $32.5 million in annual receipts for Facilities Support Services, NAICS code 561210). In these cases, we assumed that firms (and their receipts) were evenly distributed within the category and apportioned accordingly. For example, of the Facilities Support Services with receipts between $30.000 and $34.999 million, we assumed that half the firms and receipts were above the small-business threshold of $32.5 million. In other cases, data on receipts were suppressed for two or more size categories (so as to protect the confidentiality of individual firms within these categories). In these cases, we assumed that distribution of receipts in these categories (calculated as a residual of total industry receipts less receipts for known categories) reflected that of the next highest category for which receipt data for all size categories were available. For example, we used the distribution of receipts for firms classified in NAICS code 33641 to estimate the distribution of receipts for size categories with suppressed data on receipts in Aircraft Manufacturing, NAICS code 336411.

In all but one of these industries, OTSBs account for less than 14 percent of all firms. OTSBs also account for most of the receipts in all but one of these industries. This indicates a relatively high degree of industry consolidation that makes it harder for small firms to compete: A relatively small share of OTSBs conducts most of the business. In nine of the 12 industries shown, OTSBs account for more than 75 percent of industry receipts; in three of these industries, OTSBs account for more than 90 percent of industry receipts. Small businesses have the lowest prevalence in Aircraft Manufacturing, the industry in which the Air Force spends the most dollars. Increasing small-business use in these industries will likely require new strategies and significantly more effort than increasing small-business use in industries where a much smaller percentage of the revenue is with OTSBs.

The System for Award Management (SAM) is one potential way to identify small-business partners for the Air Force. Though variations in size thresholds can make it difficult to determine whether a firm is small, a great many small businesses have registered in the SAM, a necessary precursor for receiving federal contracts. Such registration indicates that they are *willing* to provide goods or services in an industry, but it does not indicate if they have ever done so, if the industry is their principal industry, or if they are *able* to meet government requirements for quality, quantity, and responsiveness.

Table 4.6 also shows the 12 leading industries in which the Air Force purchases goods and services, again ranked by the amount of goaling dollars that the Air Force spends on them and the total number of firms in these industries, according to the Economic Census. It also shows the number of total firms registering in the SAM and the number of small firms, as well as the percentage of all SAM registrants that are small businesses. As the table shows, the number of SAM registrants, both total and small business, far exceeds what the Economic Census shows in these industries. (The exceptions are Engineering Services, NAICS code 541330, and Computer Systems Design Services, NAICS code 541712.) There are two possible reasons for this discrepancy.

Table 4.6. SAM Registrants, Including Small Firms, Outnumber Total Firms Enumerated in Economic Census for Leading Air Force Industries

Industry	Total Number of Firms per Economic Census	SAM Registrants		
		Total	Small	Small as % of Total
Aircraft Manufacturing	221	2,170	1,672	77.1
Research and Development in the Physical, Engineering, and Life Sciences	11,382	28,118	18,122	64.4
Other Aircraft Parts and Auxiliary Equipment Manufacturing	770	5,215	4,118	79.0
Engineering Services	47,714	42,487	24,568	57.8
Guided Missile and Space Vehicle Manufacturing	14	875	550	62.9
Other Support Activities for Air Transportation	2,986	3,078	2,272	73.8
Search, Detection, Navigation, Guidance, Aeronautical, and Nautical System and Instrument Manufacturing	494	4,027	2,614	64.9
Nonscheduled Chartered Freight Air Transportation	206	1,373	822	59.9
All Other Professional, Scientific, and Technical Services	15,547	27,195	21,409	78.7
Facilities Support Services	1,861	15,878	11,509	72.5
Computer Systems Design Services	37,028	31,481	26,787	85.1
Wired Telecommunications Carriers	3,471	5,317	3,610	67.9

SOURCE: Special tabulations of 2007 Economic Census for SBA; SAM data.

First, a firm may register in as many as 1,000 industries, including those in which it aspires to but does not currently provide goods and services. Representatives of firms registering in the SAM may also be imprecise in noting the industries in which their firms can offer goods and services.[28]

---

[28] An additional complication in determining small businesses within an industry may be variation of size thresholds within some industries. For example, the size standard for Engineering Services (NAICS code 541330) firms is $15.0 million in average annual receipts, but firms providing Engineering Services for Military and Aerospace Equipment and Military Weapons or providing Marine Engineering and Naval Architecture have a size standard of $38.5 million. Similarly, the size threshold for firms providing Other Computer-Related Services (NAICS code 541519) is $27.5 million in revenue, but that for firms functioning as Information Technology Value-Added Resellers is 150 employees. Finally, while the size standard for R&D in the Physical, Engineering, and Life Sciences (except Biotechnology) (NAICS code 541712) is 500 employees, that for R&D firms on aircraft is 1,500 employees, for R&D firms on aircraft parts is 1,000 employees, and for R&D on space vehicles and guided missiles and related components is 1,000 employees. Such varying size standards within industries can make it difficult to accurately determine the number of small businesses available in a specific industry. This, in turn, may make it difficult to establish what may be a reasonable expectation of performance for small-business purchasing within an industry.

Second, Economic Census data on firms by industry may focus only on the principal industry in which a firm provides goods and services. Firms may provide goods and services in separate industries but perhaps may only aggregate data for the dominant industry in the Economic Census. For example, as we discuss in unpublished research conducted with our colleagues Lloyd Dixon and Aaron Kofner, firms engaged in more than one industry at one location are requested to submit separate reports on activity in each industry if their records permit such a separation. Unless firms keep their business records by Census Bureau industrial classifications rather than by the categories that they find best for their own management, this request may go unfulfilled.

The Air Force, in seeking to improve its small-business use, may wish to explore the reason for the discrepancy between SAM registrants seeking federal contracts in an industry and the number of firms that the Economic Census tabulates as providing goods and services in an industry.[29] If the SAM contains willing and able firms, then the Air Force may have a readily accessible means for expanding small-business participation. Should the SAM be inaccurate—if many firms registered in a particular industry are willing but not able to provide goods and services that meet Air Force requirements—then the Air Force may need a better count of firms, including small businesses, able to meet its requirements. Such exploration should also confirm the balance of small and other firms within industries by number and total revenue. For nearly all the above industries (excepting Guided Missile and Space Vehicle Manufacturing, NAICS code 336414), the SAM, while showing a substantial number of small businesses in these industries, also shows that such firms are a smaller percentage of all firms registered than they are of firms in the Economic Census. However, the percentage of revenue of these firms can differ vastly from their number.

---

[29] There may be similar patterns in other industries that the Air Force should explore. Small businesses are a majority of registrants in 87 percent of the industries in which the Air Force purchases goods and services. These industries accounted for 94 percent of Air Force obligations in FY 2013, as well as 95 percent of its purchases with small businesses. A small percentage of industries used by the Air Force has a low percentage of small-business registrants and represents a small percentage of obligations and an even smaller percentage of small-business use. We surmise that these are industries in which small businesses do not consider themselves to be competitive, either for federal contracts or perhaps in the broader economy.

5. Refining the Performance Expectations Methodology

Our findings point to several areas in which the Air Force could improve the accuracy of its method for setting Performance Expectations objectives, as well as some areas in which it likely cannot improve or refine the methodology's accuracy.

At the core of the current methodology are the industry and PSC combinations that the Air Force currently uses to define its markets. While these combinations do identify categories of purchases better than either industry or PSC alone, they have limitations. They may still be too broad in some circumstances, as shown by the number of combinations, each with their own level of small-business use, that cross budget categories. Assignment of NAICS codes and PSCs to contract actions can be imprecise. Air Force small-business opportunities may also be affected by broader budget category trends that the Performance Expectations methodology should perhaps consider.

While we suggest that the Air Force consider budget categories in addition to NAICS code and PSC combinations in setting performance expectations, we also suggest that the Air Force drop those markets defined by combinations of NAICS codes, PSCs, and budget categories combinations from the Performance Expectations methodology in which the Air Force cannot reasonably expect to increase its small-business use. The Air Force might, for example, drop from Performance Expectations consideration those markets in which it is already using small businesses for 90 percent or more of its purchases.

Determining small-business availability may be an additional area of difficulty for possible refinement. As noted earlier, the SAM lists firms willing to provide goods or services in an industry, although these firms may have never actually done so. The Economic Census analyzes firms in their primary industry and thus typically has fewer firms in each industry (given that many firms provide goods and services in more than one industry). Consolidation within industries is largely a result of economies of scale or scope that determine the optimal size at which firms most efficiently provide goods or services within an industry, and such changes in size may not reflect size standards for procurement preferences.[30]

SBA size standards for procurement thresholds also vary between industries and sometimes even within a specific industry, such as Engineering Services. The process that the SBA uses to set size standards includes public comment, which can lead to lowering the size standard below what the SBA's size standard methodology initially recommended.[31] This shift can affect the number of "small" firms in the marketplace with which the Air Force might contract work.

---

[30] For more on changing economies of scale or scope and optimal firm size, see Hathaway and Litan (2014).

[31] In 2011, the SBA recommended increasing the Engineering Services size standard for procurement preferences to $19 million in annual receipts for a firm. After public comments, during which some protested the higher size

The current Performance Expectations methodology uses all of DoD as a comparison group for the Air Force to determine HOT Spots. As we saw, different organizations purchase varying mixes of goods and services within different budget categories. The Performance Expectations methodology may be improved by using the Army and Navy combined as a comparison standard, or even just the Navy (its purchases by category are most similar to the Air Force), rather than all DoD.

Finally, the Air Force may want to change the terminology used in the Performance Expectations methodology to more accurately describe the opportunities in each market, replacing Accessible with More Accessible and Inaccessible with Less Accessible.

From these observations, we suggest several specific revisions for the Performance Expectations Methodology. Table 5.1 illustrates these by showing the steps in the current methodology, as listed in Table 2.1, with our recommended additions shown in bold (deletions are struck through).

Key additions and deletions include the following:

- adding "DoD budget categories" to NAICS and PSCs
- adding "Distinguish Saturated Markets" (where the Air Force has at least 90 percent small-business participation) and "Distinguish Consolidated Markets," both of which help the Air Force identify markets in which small-business growth is less likely
- deleting "in" from Inaccessible and replacing it with "less" and adding "more" to Accessible to create Less Accessible and More Accessible categories, making the category distinctions less absolute
- adding "Navy plus Army" and deleting "DoD" as the focus of identifying HOT markets.
- adding the qualifier "less the saturated-market dollars, less the consolidated-market dollars" to the 1 percent of expiring OTSB More Accessible Market not–HOT Spot dollars.

---

standard while 60 percent supported it, the standard was set at $14 million, with continued exceptions for specialties, such as Engineering Services for Military and Aerospace Equipment and Military Weapons. Using Economic Census data for this industry in 2007, we estimate that nearly 400 Engineering Services firms, with more than $5.6 billion in annual receipts, were excluded from small-business procurement preferences because of the decision to use the lower receipts threshold; see U.S. Small Business Administration, 2012.

Table 5.1. Performance Expectations Methodology and Proposed Revisions

1. Tabulate past FY DoD and Air Force goaling-dollar spending and small-business use.

2. Distinguish **More** Accessible Markets (NAICS/PSC/**budget** combinations where the Air Force has at least 1 percent small-business participation).

3. **Distinguish Saturated Markets (i.e., More Accessible NAICS/PSC/budget combinations where Air Force has at least 90 percent small-business participation).**

4. **Distinguish Consolidated Markets where small-business growth is less likely.**

5. Distinguish **Less** ~~In~~Accessible Markets (NAICS/PSC/**budget** combinations where the Air Force has less than 1 percent small-business participation).

6. Identify HOT Spots (NAICS/PSC/**budget** combinations where **Navy plus Army** ~~DoD~~ small-business use, ~~excluding the Air Force,~~ is more than double Air Force small-business use). HOT Spots can be in **More** Accessible or **Less** ~~In~~Accessible markets.

7. Identify expiring OTSB Air Force contracts.

8. Identify expiring OTSB HOT Spot and **More** Accessible minus HOT Spot contract dollars.

9. For each MAJCOM and PEO

    a. Identify past-year small-business use in dollars.

    b. Add 10 percent of expiring OTSB **More** Accessible HOT Spot dollars.

    c. Add 1 percent of expiring OTSB **More** Accessible-Market not–HOT Spot dollars, **minus the saturated-market dollars, minus the consolidated-market dollars**.

10. Set MAJCOM/PEO performance expectation to total dollars from Step 9~~7~~ divided by total past-year obligations.

The Table 5.1 suggestions appear in Figure 5.1, which is a modified version of the process discussed previously.

Figure 5.1. Proposed Performance Expectations Methodology

$51,698 million in FY2012 goaling dollars

33

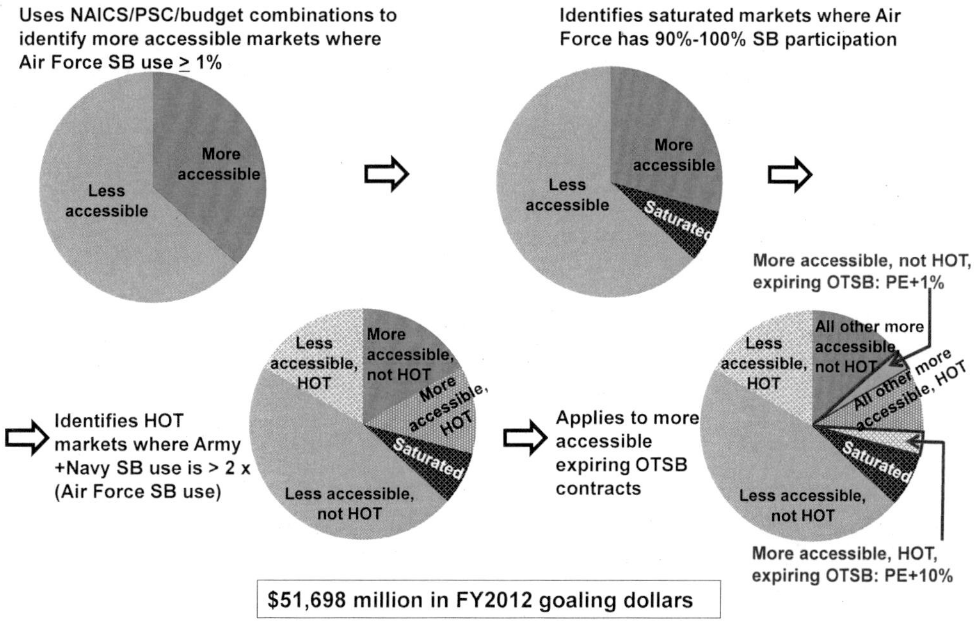

Among other changes, our recommendations would have the Air Force use the terms More Accessible and Less Accessible to describe the markets, instead of Accessible and Inaccessible, to better define the opportunities in each category. We also recommend that the methodology use industry/PSC/budget-category groupings, rather than groupings that only consider combinations of industries and PSCs, to identify More Accessible Markets. This will identify opportunities more precisely by better accounting for variation in small-business use by budget category. Within More Accessible Markets, we recommend that the Air Force identify Saturated Markets by industry/PSC/budget-category grouping; in these markets, the Air Force has at least 90 percent small-business use and therefore will find it very difficult to further expand small-business use.

We recommend that the Air Force identify OTSB expiring Consolidated Markets contracts, where small-business growth is less likely because market forces are causing firms either to increase in size or exit an industry.[32] Such markets are those where economies of scale or scope are increasing the size that a firm needs to be to compete—perhaps faster than SBA size standards recognize.

We also recommend that the Air Force identify OTSB expiring HOT Spots contracts by comparing its buying to the Army and the Navy combined, or perhaps to the Navy alone, but not to all of DoD. The services have more in common with each other in purchasing than they have with other elements of DoD. Recognizing this will help the Air Force discern areas in which it is most likely to be able to improve, as well as the most relevant lessons that others in DoD may offer.

Finally, we recommend that the Air Force subtract Saturated and highly consolidated market contracts from expiring OTSB More Accessible not–HOT Spot contracts when calculating Performance Expectations. This will help the Air Force set a more realistic and more targeted goal for the MAJCOMs and PEOs to achieve.

Data are available for many of the methodological refinements we propose, as Table 5.2 summarizes. These include data on expenditures by budget category for each industry and PSC combination, as we reviewed earlier. Identifying areas with 90 to 100 percent small-business use will enable analysts and policymakers to remove them in calculating refined Performance Expectations. The Air Force can also test different comparison groups for HOT Spots, such as Army and Navy combined or Navy alone, to discern their effects on the Performance Expectations methodology.

---

[32] While the Performance Expectations methodology focuses on OTSB contracts, when identifying opportunities to increase small-business use, MAJCOMs and PEOs may want to include expiring small-business contracts for two reasons. First, these could offer opportunities to combine some requirements from OTSB contracts into contracts amenable to small businesses. Second, expiring small-business contracts could also adversely affect small-business use if the small businesses cannot be replaced with the same or another small business.

**Table 5.2. Existing Data Allow Consideration of Many Elements
in Performance Expectations Methodology**

Area for Refinement	Action
Budget categories	Include with NAICS/PSC groups
Industries with more than 90 percent small-business use	Remove from Performance Expectations calculation
Performance measurement depending on comparison group	Test different comparison groups for HOT Spots, such as Army and Navy
Small-business availability	
Industry consolidation	Estimate number of small businesses available
SAM registration	Identify potential number of small businesses by industry
Small-business size thresholds	Tabulate by industry firms to increase small-business use

Existing data can also help identify industries that have consolidated and where small businesses may be less available and less viable over time. Such data can identify the potential number of small businesses by industry to provide goods and services to the Air Force. As earlier noted, what SAM registration indicates may sometimes be unclear (e.g., whether a firm is merely interested in providing goods and services or actually providing them), although the SAM can at least help to identify some industries with low proportions of small-business registrants. Finally, the Air Force can use a variety of sources to tabulate by industry the number of firms it may use to increase its small-business use.

Figure 5.1 illustrates the changes that would result from our recommended improvements, except for consolidation to the Performance Expectations methodology for FY 2012 goaling dollars. Rather than grouping expenditures by industry and PSC, it would do so by industry code, PSC, and budget group. A little less than 37 percent of the Air Force goaling dollars, or almost $19 billion, is in More Accessible Markets. Of that amount, nearly $4 billion (20 percent) is in Saturated Markets, with little or no room for improvement. More than $6 billion (almost 34 percent) of Accessible Markets are HOT Spots.

The focus of the Performance Expectations methodology is on all expiring contracts. In the markets that are More Accessible, not HOT Spots, and also not Saturated, there are almost $9 billion in obligations. Seventeen percent of these are expiring (more than $1.5 billion) and have a Performance Expectations increase of 1 percent. In markets that are More Accessible and HOT Spots but not Saturated, there are more than $6 billion in obligations. Twenty-seven percent of these are expiring (more than $1.7 billion) and have a Performance Expectations increase of 10 percent. While still giving special attention to expiring OTSB contracts, the recommended methodology would focus more attention on markets where small-business use is less prevalent and still able to grow.

We used FY 2012 FPDS-NG data to illustrate our proposed changes to the Performance Expectations methodology.[33] Table 5.3 shows the Performance Expectations for each MAJCOM (the goals for their small-business use, shown in percentages) and summarizes the comparisons resulting from our recommendations. (As before, MAJCOMS are listed in the order of their percentage of Air Force purchases.) The column on the left shaded in yellow is MAJCOM Performance Expectations using the current Air Force methodology. The column on the right shaded in light green provides Performance Expectations calculated with recommended refinements to the methodology: creating HOT Spots based on Army plus Navy small-business use, forming NAICS/PSC/budget groupings, and eliminating saturated groupings.

Table 5.3. Eliminating Saturated Markets Would Have the Greatest Effect on Performance Expectations Calculations for MAJCOMs

MAJCOM (Col. 1)	NAICS/PSC Group and HOT Spot Comparison		NAICS/PSC/Budget Group and HOT Spot Comparison		NAICS/PSC/Budget Minus Saturated Market and HOT Spot Comparison	
	DoD Minus Air Force (Col. 2)	Army and Navy (Col. 3)	DoD Minus Air Force (Col. 4)	Army and Navy (Col. 5)	DoD Minus Air Force (Col. 6)	Army and Navy (Col. 7)
AFMC	10.89	10.90	10.97	10.98	10.97	10.98
AFSPC	6.47	6.47	6.47	6.45	6.47	6.45
ACC	57.48	57.49	57.49	57.56	57.48	57.56
AETC	51.81	51.81	51.93	51.98	51.93	51.97
AMC	68.31	68.32	68.43	68.56	68.43	68.56
PACAF	50.82	50.83	50.78	50.90	50.78	50.90
AFDW	26.07	26.15	26.17	26.18	26.17	26.18
AFISRA	26.94	27.44	26.91	27.50	26.91	27.50
AFGSC	75.45	75.45	75.41	75.74	75.40	75.73
USAFA	75.22	75.25	75.22	75.32	75.22	75.32
AFRC	69.73	69.72	69.72	69.76	69.70	69.73
AFSOC	71.03	71.03	71.00	71.15	70.99	71.15
USAFE	20.13	20.13	20.08	23.58	20.08	23.57
AFOTEC	59.74	59.74	59.74	59.74	59.73	59.73
Air Force–wide	15.04	15.06	15.10	15.13	15.10	15.13

SOURCE: FY 2012 FPDS-NG DoD goaling dollars.

---

[33] Another possible consideration is a comparison between unconsolidated groupings (e.g., those where small businesses represent 5 percent or more of an industry's revenue) to all groupings. Because of time and resource constraints, we were unable to explore this area.

Employing small-business use by the Army plus Navy rather than by DoD (exclusive of the Air Force) to determine HOT Spots made very little difference in MAJCOM Performance Expectations, as shown by comparing column two with column three. Of the 14 MAJCOMs listed, six stayed the same, five changed by only 0.01 percentage point, and three changed by 0.03 to 0.50 percentage points.[34] The changes were caused by some Accessible not–HOT Spot markets (for example, more than $32 million in NAICS 443120, Computer and Software stores/PSC 7030, ADP Software) shifting to HOT Spot markets that have a higher goal, while some other Accessible HOT Spots shifted to not–HOT Spots with a corresponding lower goal.

Using NAICS/PSC/budget groups instead of using NAICS/PSC groups also made a small difference, as shown by comparing column two to column four (five went up by 0.01 to 0.12 percentage points, six went down by 0.01 to 0.05 percentage points, and three stayed the same) and column three to column five (12 went up by 0.03 to 0.45 percentage points, one went down by 0.02 percentage points, and another stayed the same).[35] The overall Air Force change from the current methodology was 0.09 when comparing the Air Force to the Army and Navy alone, rather than the entire rest of DoD. Many of the changes that did occur resulted from categories shifting between HOT Spot designations, including, for example, more than $320 million in NAICS 541330, Engineering Services/PSC R449, Support—Professional: Other/O&M that shifted from a not–HOT Spot to a HOT Spot designation.

Finally, five MAJCOM Performance Expectations fell slightly (none by more than 0.03 percentage points) when we eliminated saturated groupings (such as NAICS 236220, Commercial and Institutional Building Construction/PSC 2212, Repair or Alteration of Miscellaneous Buildings) from the Performance Expectation calculation, while the rest stayed the same, including that for the Air Force as a whole, as shown by comparing column seven to column five. Although eliminating saturated groupings by itself did not create substantial change for most MAJCOMs, we did see several Performance Expectation calculations increase from that calculated by the current method. As discussed earlier, this occurs because some of the remaining NAICS/PSC/budget combinations switched from not HOT to HOT. Nevertheless,

---

[34] It seems counterintuitive that some Air Force Performance Expectations would go up when the comparison group shrinks. This occurs because we are comparing percentages, not absolute numbers for specific NAICS/PSC combinations. When the comparison group shrinks, the denominator may shrink more than the numerator if the groups removed have a lower small-business use rate than those that remain; therefore, the comparison percentage can increase and the comparison can shift from not HOT to HOT. Further, when comparing percentages, total and small-business spending may be small in one comparison group, but the percentage is high because of their relative value. Total and small-business spending could be much higher in the other comparison group, but the percentage is much smaller. For a specific NAICS/PSC combination, this can lead to low total spending in one comparison group driving a larger Performance Expectation increase for a relatively higher total spending in another comparison group and vice versa. Lastly, Performance Expectations are more likely to change more for organizations that spend less than for those that spend more because a change in status from not HOT to HOT and vice versa in combinations with relatively large spending has a bigger effect on a smaller overall total spending.

[35] By adding budget categories, we make the comparison more realistic, but shrinking the sizes of the comparison groups can further change what is HOT and not HOT.

subtracting consolidated industries from the groupings is likely to further reduce some MAJCOM Performance Expectations, particularly those with substantial PEO obligations.

The combined effects for MAJCOM Performance Expectations is small—less than 0.2 percentage points in most cases. Because the purpose of the recommendations is to increase the methodology's accuracy and not to increase or decrease the goals themselves, it is not surprising that the goals change little in most cases for the Air Force overall and for most MAJCOMs,

However, in some cases, even though the change is small in percentage, it is quite large in relative terms. The Performance Expectation for U.S. Air Forces in Europe (USAFE) rose 3.44 percentage points, but this is an increase of 17 percent over the original Performance Expectation of 20.13 (i.e., 3.44 percentage points is 17 percent of 20.13). This relatively large increase is the result of one expiring contract moving from the not HOT to the HOT category with the recommendations.

Comparing the recommended Performance Expectations to those currently used for PEOs showed the effects of both adding budget categories and dropping Saturated Markets, though these were generally more modest than the effects seen for MAJCOMs. Table 5.4 shows results for 13 PEOs. (As before, PEOs are listed in the order of their share of Air Force purchases, although the Space PEO, because it did not exist in FY 2012, is not listed.) The line at the bottom of the table is for non-PEO expenditures and offers a comparison point for PEO expenditures.

Using small-business use by the Army and Navy rather than by DoD as a whole (exclusive of the Air Force) to determine HOT Spots did not change the Performance Expectation for nine PEOs. Of the remaining five PEOs, the Performance Expectation went down by 0.01 or 0.02 percentage points for four and went up by 0.42 percentage points for one. Meanwhile, the non-PEO went up by 0.02.

Using NAICS/PSC/budget groupings instead of using NAICS/PSC groupings made a small difference, as shown in Table 5.4 by comparing column three to column five. The Performance Expectations for three PEOs stayed the same, while those for two PEOs went up by 0.24 to 0.49 percentage points and those for eight PEOs went down from 0.02 to 0.46 percentage points. The Performance Expectation for the non-PEO went up by 0.09 percentage points.

When we eliminated saturated groupings, as shown by comparing columns five to seven, none of the PEOs' Performance Expectations changed, while that for the non-PEO went down by 0.01 percentage points. This is not surprising, given that the vast majority of PEO dollars are spent on weapon systems, which are highly unlikely to be in markets that are saturated. We would, however, expect that PEO Performance Expectations would decrease when Consolidated Markets were eliminated.

Thus, a revised methodology that uses budget categories and an Army-plus-Navy comparison while excluding Saturated Markets would lead to a lower Performance Expectation in most PEOs, though most of these reductions are modest. Eight PEO Performance Expectations fell from 0.02 to 0.11 percentage points, three stayed the same, and two rose from 0.24 to 0.50 percentage points, while the non-PEO Performance Expectation rose by 0.10 percentage points.

However, as with the MAJCOMs, there are cases in which small percentage increases translate into large relative changes. Among the PEOs, the Performance Expectations revision would lead to an increase of 0.5 percentage points for ISR and SOF, but because its Performance Expectation was only 2.6 percent at the start, this translates into an increase of 19 percent (i.e., 0.5 percentage points is 19 percent of 2.6). Similarly, a revised methodology would reduce the Performance Expectation for Space Launch by 0.11 percentage points, from 0.16 to 0.05. This decrease reflects a relative drop of 68 percent (i.e., 0.11 percent is 68 percent of 0.16). Both of these changes are the result of NAICS/PSC/budget combinations shifting between the not-HOT and HOT categories under the recommendations. For ISR and SOF, one contract was in the not-HOT category under the original calculation and was in the HOT category under the recommendations. For Space Launch, two contracts shifted from the HOT category to the not-HOT category under the recommendations.

In practice, these large changes, and the one described above for the USAFE MAJCOM, indicate a variation in the Performance Expectations that is at least sometimes so great that it might make planning difficult at best for these organizations. When the large change is an increase, it may simply be too large for a MAJCOM or PEO to reasonably meet. However, the reason for this variation, and the resulting solution to it, indicate that such variation is not likely to pose such a problem. In any given year, a large OTSB Air Force contract can expire, and the Performance Expectation could rise substantially. With the recommendations here, comparing the original Performance Expectation with the recommended one creates a similar possibility: The revised Performance Expectation might be substantially lower or higher than the original because under the recommendations, a large expiring Air Force contract moves from the not-HOT category to the HOT category or vice versa.

The solution to this for calculating yearly changes in Performance Expectations is to pool the previous years of comparison in order to smooth outlier Air Force contracts, such as those that affected the Performance Expectations for USAFE, ISR and SOF, and Space Launch. Rather than tabulating only the past FY DoD and Air Force goaling-dollar spending and small-business use in Step 1 of the methodology, pooling prior years would tabulate and average the past two or three years of spending and small-business use. This would smooth the effects of outlier Air Force contracts across two or three years.[36] In addition, SAF/SB could check for anomalies where the Performance Expectation emerges unusually higher or lower than the previous year, even after such smoothing, and determine whether further adjustments were needed (i.e., whether it was practically realistic to expect that the spending on the expiring OTSB contract might reasonably convert into some new small-business spending).

Finally, once initial programming is set up, we recommend pooling multiple years of data and repeating the analysis over time to gain a more accurate assessment of the effects of the changes proposed here. By definition, the Performance Expectations methodology is an applied

---

[36] Unfortunately, data limitations prevented us from demonstrating this effect.

measure, and repeated application will clarify which of these changes demonstrate consistent effects that may want to be applied in other contexts.

Table 5.4. Eliminating Saturated Markets and Using Army-Plus-Navy Comparison Would Slightly Affect Performance Expectations Calculations for PEOs

	Performance Expectation Calculation					
	NAICS/PSC Group and HOT Spot Comparison		NAICS/PSC/Budget Group and HOT Spot Comparison		NAICS/PSC/Budget Minus Saturated Market and HOT Spot Comparison	
PEO (Col. One)	DoD Minus Air Force (Col. Two)	Army and Navy (Col. Three)	DoD Minus Air Force (Col. Four)	Army and Navy (Col. Five)	DoD Minus Air Force (Col. Six)	Army and Navy (Col. Seven)
Space Systems	3.64	3.64	3.64	3.64	3.64	3.64
Mobility	1.05	1.04	1.01	1.01	1.01	1.01
ISR and SOF	2.60	2.61	3.09	3.10	3.09	3.10
Fighters/ Bombers	1.03	1.03	1.01	1.01	1.01	1.01
Battle Management	4.00	4.00	3.94	3.97	3.94	3.97
Tankers	0.05	0.05	0.07	0.05	0.07	0.05
Agile Combat Support	10.62	10.60	10.56	10.57	10.55	10.57
Space Launch	0.16	0.16	0.15	0.05	0.15	0.05
Weapons	4.45	4.45	4.69	4.69	4.69	4.69
C3I&N	5.91	5.91	5.87	5.87	5.87	5.87
Strategic Systems	1.45	1.45	1.43	1.43	1.43	1.43
B&ES	25.57	25.99	25.52	25.53	25.52	25.53
CM Support	9.27	9.27	9.27	9.27	9.27	9.27
Non-PEO	37.05	37.07	37.08	37.16	37.08	37.15

SOURCE: FY 2012 FPDS-NG DoD goaling dollars.

6. Conclusions

As the Air Force seeks to refine its Performance Expectations methodology, we recommend that it consider the findings and recommendations described in this report as ways to increase methodology accuracy.

First, Air Force spending by MAJCOM and PEO has been relatively consistent across time. Small-business use has often been greater for those MAJCOMs and PEOs that spend fewer dollars and presumably have requirements more amenable to small-business purchases. By tabulating past purchases by MAJCOM and PEO, effectively standardizing future expectations to them, the current Performance Expectations methodology recognizes this pattern, and any effort to refine it should do the same.

Second, efforts to increase small-business spending by MAJCOMs and PEOs should consider differences among them and with other services. As noted, the amenability of each MAJCOM and PEO to small-business purchases varies. Beyond that, Air Force purchases differ from those of other parts of DoD. While the Navy and the Army may offer some relevant comparisons and lessons for Air Force efforts to increase small-business use, other elements of DoD, such as DLA, are less relevant.

Third, small-business spending differs by industry, PSC, and budget category. The current Performance Expectations methodology recognizes this in part by considering industry and PSC combinations. Nevertheless, as our work shows, it also needs to consider budget categories. A large group of industry/PSC/budget-category combinations accounts for only about 7 percent of total Air Force obligations but most of its small-business obligations. Recognizing this in the Performance Expectations methodology will help focus efforts where the Air Force might make gains.[37]

Fourth, extant data can offer many insights but can be improved in some cases. These include narrowing industries that have multiple size standards (different numbers of employees or different amounts of revenue for different parts of the same industry) for what constitutes a small business or span multiple budget categories, such as Procurement and O&M. Extant data on industry/PSC/budget-category combinations shows where the Air Force might best focus its efforts to improve small-business use, away from categories where small businesses already saturate the Air Force market. SAM and Economic Census data can offer insights on broader possibilities for the Air Force in improving small-business use, identifying where small businesses are plentiful or few. Yet data may have some shortcomings, as evident in the

---

[37] The appendix presents further discussion on projecting small-business use as spending by budget categories changes.

discrepancies between the SAM and the Economic Census data, as well as in questions regarding whether these sources include too many or too few firms.

Finally, we note that the Performance Expectations methodology addresses only one aspect of small-business use, determining and setting accurate goals for that use. Continuing to monitor and refine the accuracy of this methodology will help this tool remain useful as the Air Force seeks to increase small-business use. Pooling the last two to three years of data, rather than using only the previous year, is one example of such monitoring and refining. The Air Force will also need to continue its ongoing efforts to examine other aspects of its purchasing that affect small-business use, including breaking out components from large weapon systems.

Appendix

How to Identify Consolidated Markets

Consolidated Markets are best identified with the special tabulations of the Economic Census that the Census Bureau does for the SBA on firms and receipts by number-of-employee or size-of-receipts categories. Missing size-category data for these can be imputed through the method of Reardon and Moore (2005). This method assumes that within a missing category, the distribution of firms or revenue is linear and in proportion to the next-highest grouping in the NAICS hierarchy. Once missing data in published industry distributions of businesses by size have been imputed, the number of firms and their revenue above and below the size standard can be calculated. From these calculations of firms and their revenue above and below the size standard and an agreed-upon consolidation threshold (e.g., at least 95 percent of industry revenues with OTSBs), consolidated industries can be identified. We recommend eliminating such industries from performance-expectation calculations in our revised methodology.

How to Estimate Future Air Force Small-Business Use from Past Use and Budget Projections

Estimating future small-business use begins with using the Treasury Account Symbols on FPDS-NG goaling-dollar contract action reports to group contract actions into budget categories. The percentage of each budget category that was spent with small businesses is then calculated. These calculations should be applied to the latest Air Force budget-category projections, as published in the DoD Green Book (Office of the Deputy Under Secretary of Defense [Comptroller], 2015). Future small-budget use is calculated by multiplying the percentage of small-business use for the last FY by total dollars in the budget category to estimate total small-business dollars in that budget category, then by summing projected small-business dollars across budget categories, then by dividing the total projected budget dollars for the future year to obtain a projected small-business use for the future year.

Acknowledgments

We wish to thank Mark S. Teskey, director; Joseph M. McDade, Jr., former director; Carol E. White, deputy director; and Brett T. Scheideman, Small Business Acquisition Portfolio manager, Air Force Small Business Programs, for their guidance and support of this project. We thank our former RAND colleague, Michael R. Thirtle, for his help launching this project. We also thank Laura Baldwin and Jim Powers of RAND for their careful reviews of the document.

Abbreviations

ACC	Air Combat Command
AETC	Air Education and Training Command
AFDW	Air Force District of Washington
AFGSC	Air Force Global Strike Command
AFISRA	Air Force Intelligence, Surveillance, and Reconnaissance Agency
AFMC	Air Force Materiel Command
AFOTEC	Air Force Operational Test and Evaluation Center
AFRC	Air Force Reserve Command
AFSOC	Air Force Special Operations Command
AFSPC	Air Force Space Command
AMC	Air Mobility Command
B&ES	Business and Enterprise Systems
C3I&N	command, control, communications, intelligence, and networks
CM	Combat and Mission
DLA	Defense Logistics Agency
DoD	U.S. Department of Defense
FPDS-NG	Federal Procurement Data System–Next Generation
FY	fiscal year
HOT	High-Opportunity Target
ISR	intelligence, surveillance, and reconnaissance
MAJCOM	major command
NAICS	North American Industry Classification System
O&M	Operations and Maintenance
OCE	Office of the Corps of Engineers
ODA	other defense agencies
OTSB	other-than-small business

PACAF	Pacific Air Forces
PEO	Program Executive Office
PSC	Product or Service Code
RDT&E	Research, Development, Test, and Evaluation
SAF/SB	Air Force Office of Small Business Programs
SAM	System for Award Management
SB	small business
SBA	Small Business Administration
SOF	Special Operations Forces
USAFA	United States Air Force Academy
USAFE	United States Air Forces in Europe

References

Carter, Ashton B., "Remarks at National Press Club," Washington, D.C., May 7, 2013. As of June 14, 2014:
http://archive.defense.gov/Speeches/Speech.aspx?SpeechID=1775

Deans, Graeme K., Fritz Kroeger, and Stefan Zeisel, "The Consolidation Curve," *Harvard Business Review*, Vol. 80, No. 12, December 2002. As of January 12, 2017:
https://hbr.org/2002/12/the-consolidation-curve

Federal Procurement Data System—Next Generation, "Small Business Goaling Report, Fiscal Year: 2014," undated. As of July 7, 2015:
https://www.fpds.gov/downloads/top_requests/FPDSNG_SB_Goaling_FY_2014.pdf

Gates, Robert M., "Remarks by Secretary Gates to the American Enterprise Institute, Washington, D.C.," May 24, 2011. As of June 14, 2014:
http://archive.defense.gov/Transcripts/Transcript.aspx?TranscriptID=4827

Hathaway, Ian, and Robert Litan, "The Other Aging of America: The Increasing Dominance of Older Firms," Washington, D.C.: The Brookings Institution, July 2014. As of August 5, 2015:
https://www.brookings.edu/wp-content/uploads/2016/06/
other_aging_america_dominance_older_firms_hathaway_litan.pdf

McDade, James T., "Small Business Program Update," Air Force Office of Small Business Programs, Department of the Air Force, March 11, 2013. As of February 22, 2016:
http://www.ndia.org/Divisions/Divisions/SmallBusiness/Documents/
McDade_20130311_Triad%20Brief_Final.pdf

Moore, Nancy Y., Clifford A. Grammich, and Judith D. Mele, *Small Business and Strategic Sourcing: Lessons from Past Research and Current Data*, Santa Monica, Calif.: RAND Corporation, RR-410-OSD, 2014. As of May 15, 2015:
http://www.rand.org/pubs/research_reports/RR410

Office of the Deputy Under Secretary of Defense (Comptroller), *National Defense Budget Estimates for FY 2016*, Washington, D.C., March 2015. As of May 18, 2015:
http://comptroller.defense.gov/Portals/45/Documents/defbudget/fy2016/
FY16_Green_Book.pdf.

Porter, Michael E., *Competitive Advantage: Creating and Sustaining Superior Performance*, New York: The Free Press, 1985.

Public Law 85-536, Small Business Act, as amended, January 3, 2013. As of August 7, 2015:
https://www.sba.gov/sites/default/files/policy_regulations/Small%20Business%20Act_0.pdf

Reardon, Elaine, and Nancy Y. Moore, *The Department of Defense and Its Use of Small Businesses: An Economic and Industry Analysis*, Santa Monica, Calif.: RAND Corporation, DB-478-OSD, 2005. As of May 15, 2015:
http://www.rand.org/pubs/documented_briefings/DB478

Stackley, Sean J., "Meeting Small Business Goals in FY 2013," Assistant Secretary of the Navy for Research, Development and Acquisition, Department of the Navy, December 13, 2012. As of February 22, 2016:
http://www.secnav.navy.mil/rda/Policy/2012%20Policy%20Memoranda/
13dec2012asnrdasbgoalsmemo.pdf

U.S. Department of the Army Office of Small Business Programs, "Newsletter," December 2013. As of February 22, 2016:
http://www.sellingtoarmy.info/sites/default/files/newsletter/
ArmyOSBP_Newsletter_December2013.pdf

U.S. Department of Defense Office of Small Business Programs, "Small Business Goaling Exclusions," undated a. As of November 29, 2016:
http://www.acq.osd.mil/osbp/gov/goalingExclusions.shtml

———, "Small Business Maximum Practicable (MaxPrac) Opportunity Analysis Model," undated b. As of August 5, 2015:
http://www.acq.osd.mil/osbp/gov/index.shtml#MaxPrac

———, "Small Business Program Goals," undated c. As of February 22, 2016:
http://www.acq.osd.mil/osbp/statistics/sbProgramGoals.shtml

U.S. Small Business Administration, "Small Business Size Standards: Professional, Technical, and Scientific Services," *Federal Register*, Vol. 77, No. 28, February 10, 2012, pp. 7490–7515.

———, "Table of Small Business Size Standards Matched to North American Industry Classification System Codes," July 14, 2014. As of November 7, 2014:
https://www.sba.gov/content/small-business-size-standards

Wright, Marcie C., "Secretary of Defense Discusses Budget Cuts, Changing Benefits at Fort Bliss Town Hall," Army.mil, January 17, 2012. As of November 29, 2016:
https://www.army.mil/article/72041/
Secretary_of_Defense_discusses_budget_cuts__changing_benefits_at_Fort_Bliss_town_hall